KU-527-861

Early Man and the Cosmos

Evan Hadingham

Early Man

Also by Evan Hadingham

The Fighting Triplanes
Ancient Carvings in Britain
Circles and Standing Stones
Secrets of the Ice Age

and the Cosmos

William Heinemann, Ltd.  London

LIBRARY
W. R. TUSON COLLEGE
PRESTON.

William Heinemann Ltd
10 Upper Grosvenor Street, London W1X 9PA

LONDON MELBOURNE TORONTO
JOHANNESBURG AUCKLAND

First published in Great Britain 1983

W· R·TUSON

© 1983 by Evan Hadingham

SBN 434 31108 1

Printed in the United States of America

Author's Note

This book explores the observations, myths and outlook of prehistoric skywatchers, and of more recent "primitive" astronomers who had little or no contact with our scientific tradition. As an archaeologist seeking to piece together the fragments of vanished cultures, I am interested in how these observers used their knowledge of the sky in their calendars, ritual beliefs and political purposes. This book is not a textbook on astronomy, and therefore I have deliberately kept the discussion of our modern understanding of astronomical events to a minimum.

I have also avoided the use of technical terms. Even the words "archaeoastronomy" or "astroarchaeology" have been omitted here, although they are widely used to refer to the study of ancient astronomy. This is partly for the sake of simplicity, and also because these words may suggest that we can examine astronomical lore separately from the total pattern of beliefs and values of a vanished people. In fact, the main purpose of this book is to demonstrate the remarkable connections that existed between the observations of the skywatchers and the sacred and everyday aspects of their lives.

Most of the dates cited in the text are based on the radiocarbon technique (explained in the Glossary on p. 254). These dates are only rough guides to the true age of the object or structure in question, and could be in error by several centuries. The reader who is aware of the complex scientific controversy over the accuracy of radiocarbon dates should note that all dates are given here, approximately, in their corrected (BC) form.

This book is the result of fifteen years' interest in ancient astronomy. Obviously, it would be difficult to list all those who served to inspire and guide my studies, but special thanks must be recorded here to the following individuals:

To Anthony Aveni, for his encouragement and good humor in the course of two Colgate University research expeditions in Central America. My participation in the fieldwork was an invaluable introduction to the world of the ancient Maya.

To Aubrey Burl and Alex Hooper, for their perseverance in assisting me with field investigations of standing stones in Brittany.

To Gerald Hawkins, for his advice over many years, and for the opportunity to examine astronomical sites in Scotland during two expeditions led by him and organized by Earthwatch, Belmont, Mass.

To Alexander and Archie Thom, who have always assisted me despite my criticisms of their theories. All those concerned with ancient astronomy owe an immense debt to their hard work and dedication.

To Ray Williamson, who awoke my interest in the cultures of

the American Southwest and sustained it with his generous guidance.

To my parents and to Janet Johnson for their constant encouragement and editorial advice.

A grant from the International Astronomical Union enabled me to attend the Oxford Archaeoastronomy Conference in 1981, and my thanks go to its organizers, Douglas Heggie and Michael Hoskin.

I am grateful to the following for their contributions and criticisms: Anne Atheling, John Carlson, Von Del Chamberlain, Dave Drucker, Jack Eddy, Andrew Fleming, Owen Gingerich, Gary Gossen, Horst Hartung, Ken Hedges, Douglas Heggie, Travis Hudson, Bill Hyder, Ed Krupp, Susanne Marcus, Mark Matthewman, Steve McCluskey, Jon Patrick, Michael Roberts and Anna Sofaer.

Finally, my thanks go to the staff of the Tozzer Library, Peabody Museum, Harvard University, where I researched this book.

Evan Hadingham
Cambridge, Mass., 1983

This book is for Janet,
with my love and thanks.

Contents

Author's Note		v
Foreword		3

PART I:
INTRODUCTION

1. Ancient Astronomy and the Roots of Science	10
Our Debt to the Babylonian Astronomers	10
Soothsayers and Skeptics	13
Time and the Pyramid: The Legacy of Egypt	19
Sky Visions, Ancient and Modern	24

PART II:
THE SKY AND
THE STONES

2. Superstition or Science?	33
Megalithic Masterminds	33
A Golden Age or a "Menacing World"?	34
The Geometry of Stone Circles	40

3. Stonehenge Reconsidered	42
A Midsummer Mystery	42
The Errors of Stonehenge	46

4. Death and the Sun in Ancient Britain	50
Newgrange: The Sun Reborn	50
A Doorway to the Dead	53
The Cairn of the Ancestors	57

5. The Moon and the Megaliths	61
The Lunar Rhythm	61
Moon Rites in Ancient Scotland	63
Did They Predict Eclipses?	67

6. The Riddle of the Fairy Stone	74
The Great Broken Stone of Brittany	74
Cows, Stones, and Sex: The Meaning of the Breton Megaliths	77

PART III:
THE SKY PRIESTS OF
NORTH AMERICA

7. The Hunters and the Heavens	84
The Ice Age Moonwatchers	84
Who Were the First Americans?	88
The Shaman's Sky Visions	92

8. The Counting of the Moons   96
    The Flow of Time   96
    The Moon Calendars of North America   101
    The Pawnee, Star People of the Plains   105

9. "Crystals in the Sky": The Astronomy of the Chumash   110
    The Chumash, Californian Artists and Astronomers   110
    A Power Struggle in the Cosmos   114
    The Cults of California   120

10. The Sun Priests of the Southwest   125
    The Pueblo Sunwatchers   125
    "The Ways of Our Fathers"   127
    The Year of the Hopi: "Bound Up in Time"   133
    Hopi Priests and Planters   136

11. The Astronomy of the "Old Ones"   141
    The Star, the Moon, and the Sacred Hand   141
    The Sun Towers of Hovenweep   145
    Daggers of Light: The Sun Shrine Discoveries   147
    The Chaco Achievement   156
    The Fate of the "Old Ones"   160

12. Under the Tropical Sun   165
    The Roof of Voyaging   165
    Sun Lines in Ancient Peru   170
    Conquest and the Calendar   176
    Seeking the Tropic: The Carved Circles of Mexico   181

13. The Hidden Lines of Uxmal   186
    The Palace of Venus   186
    Secrets of the Nunnery   190

14. Beans and Crystals: The Rites of the Modern Maya   194
    The Code of the Chamulas   194
    The Sacred Count   198
    The Sun Stones of the Maya   201

15. Lords of Palenque   204
    The Crypt in the Pyramid   204
    In the Jaws of the Sky Dragon   208
    The Legacy of Lord Pacal   211
    Mayan Myths and "Magic Numbers"   216

PART IV: SKYWATCHERS OF THE TROPICS: THE ANCIENT AND MODERN MAYA

16. **Prophecy and Precision: The Written Evidence**
    for Mayan Astronomy   219
    The Burning of the Books   219
    Mayan Methods of Prediction   221
    The Super-number of Venus   224
    Towers of the Venus God   226

PART V: 17. **From the Maya to the Megaliths**   234
CONCLUSION   Were There "Wise Men" in Wessex?   234
                   Revealing the Maya World   239
                   Chiefs, Priests, and Calendars   242
                   The Unifying Vision   245

        Glossary   251

        Bibliography   257

        Sources of Quotations   269

        Photo Credits   272

        Index   273

CHRONOLOGICAL TABLE

BC	EUROPE	AMERICAS	ASIA	BC
6500			Farming under way in Asia	6500
6000	End of Channel land bridge: Britain now an island	First attempts at agriculture in Mexican highlands		6000
5500				5500
5000	Early Neolithic: first farmers and megalithic tombs in Britain and Brittany	Cultivation of corn in Mexico		5000
4500			Copper smelting and casting in Mesopotamia	4500
4000			Pottery kilns, invention of wheel in Mesopotamia	4000
3500	Building of Newgrange, Ireland		Invention of writing in Mesopotamia	3500
3000	Stone circles, 3500-1200 Earliest construction at Stonehenge	Pottery appears in Mexico, Colombia, Ecuador	Hieroglyphic writing in Egypt	3000
2500	Maes Howe		Pyramids in Egypt, Chinese invent potter's wheel	2500
2000	Final phase of Stonehenge	First permanent villages in Mesoamerica	First Babylonian ziggurats	2000
1500	Bronze introduced to Britain	Earliest known Mesoamerican monuments and temples, growth of Olmec civilization on Gulf Coast	Old Babylonian period, empire of Hammurabi, 1792–1750	1500
1000	Collapse of Minoan Empire Trojan War		Chinese record eclipses from 1400	1000
500	Greek alphabet 624 Birth of Thales	Monte Albán founded	Seleucid Empire, climax of Babylonian learning	500
BC	Greek science begins	First recorded dates in Mesoamerica, pottery in	Chinese record sunspots from	**BC**
AD 100	Roman Empire (1st-5th c) 140 Ptolemy's Almagest	North American Southwest	28 BC	**AD** 100
200		Teotihuacán Empire, Central Mexico	Unification of China	200
300		Nazca culture, Peru	Water-powered astronomical globes	300
400	Huns invade Europe	Classic Maya civilization (4th–10th c)		400
500				500
600	Angles and Saxons control England	Collapse of Teotihuacán		600
700		Climax of Maya architecture and learning	Beginning of Muslim conquests	700
800	Charlemagne founds Holy Roman Empire	Collapse of Lowland Maya Building of Uxmal and other sites in Yucatán	Flowering of Arab science at Baghdad	800
900			Oldest existing Chinese star map	900
1000	Vikings discover North America	Caracol at Chichén Itzá, building of Chaco Canyon towns		1000
1100			Su Sung builds astronomical clock tower, 1088	1100
1200	Crusades			1200
1300			Chinese invent equatorial mounting for astronomical instruments	1300
1400	Beginning of Renaissance	Aztec Empire, Mexico		1400
1500	1473 Birth of Copernicus	Inca Empire, Peru 1519 Cortez lands in Mexico		1500
1600	1543 Publication of De Revolutionibus by Copernicus			1600

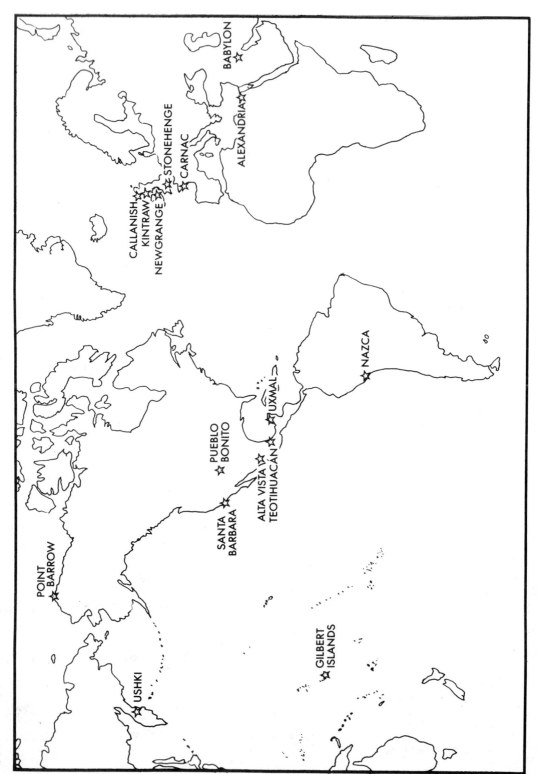

Map of some of the important sites discussed in this book.

Foreword

Over twenty-five hundred years ago, immense manmade towers of mud brick rose up toward the night skies of Mesopotamia. Here, in the "land between the rivers" of present-day southern Iraq, the towers, or "ziggurats," must have provided perfect observation points for the scribes who had developed an unrivaled knowledge of the sky. Without the aid of telescopes or mechanical devices, they predicted the movements of the sun, moon, and planets more accurately than scientists of the Victorian age.

The fame of the temple-towers spread throughout the ancient world. For the Jews held captive in the city of Babylon, the ziggurats were symbols of the arrogance and ambition of the

A stairway to the moon god. The approach to the restored ziggurat of Ur in southeast Iraq, believed to date in its earliest stages to about 2100 B.C. This was the view of the Babylonian priests as they began the long climb to the shrine on the summit.

heathen, and this passed into Old Testament tradition with the story of the Tower of Babel, "whose top may reach unto heaven."

For the Babylonians themselves, the ziggurat with its steep exterior stairway represented the link between mankind and the gods. The great king Nabopolassar—father of the Nebuchadnezzar who is familiar to us from the Bible—ruled from 625–605 B.C. In an inscription he tells how the hero-creator god Marduk instructed him to restore the tower of Babylon, "which had become weakened by time and fallen into disrepair; he commanded me to ground its base securely on the breast of the underworld, while its pinnacles should reach upwards to the skies."[1]

The renovation of the tower must have involved a huge labor force. When Alexander the Great decided to demolish its dilapidated remains three centuries later, he is said to have hired ten thousand workmen for two months to accomplish the task. The only trace of the ziggurat uncovered during excavations around the turn of the century was its foundation, nearly three hundred feet square. The casing of the tower, composed of mud bricks cemented with bitumen, was forty-five feet thick.

This great structure connected the Babylonians to the divine order of the heavens. They explained events in the sky in myths describing the actions of their gods, yet they did not neglect the human intellectual skills involved in observing them. By the time of Alexander's conquests, these skills had advanced to a remarkable level. For instance, it had become customary for the Babylonians to measure the interval between the start of an eclipse and either sunrise or sunset with the aid of a water clock. The various stages of the eclipse were also duly observed and timed. About forty eclipses are recorded on clay tablets with such accuracy and detail that scientists today regard the information as a vital clue in investigating modern problems of astrophysics, such as the changing speed of the earth's rotation. One Babylonian estimate of the exact length of the year is less than half an hour longer than the figure established by modern astronomers.

How were such achievements possible two thousand years before Galileo turned the first telescope toward the planets? The accomplishments of the ancient world—the erection of mighty monu-

The ziggurat of Ur formed part of a precinct of palaces and priestly residences sacred to the moon god Nanna-Sin. The moon god's meals were prepared in a kitchen on the northwest side of the ziggurat.

ments like the ziggurats, or the intricate calculations of eclipses preserved on the clay tablets—seem inconceivable to us in the absence of modern-day machines and instruments.

Indeed, it is hard for us to imagine the impact of the night sky on the minds of prehistoric people. In modern cities we are hemmed in by tall buildings, and street lighting overpowers the stars. Even the people of a century ago had more awareness of the sky than we do now. What was it like, then, for the skywatchers of the ancient past, who did not have our knowledge of the physical universe to explain away the stars? To them, the night sky was a source of wonder and mystery. The changing appearances and movements of the sun, moon, stars, and planets formed a complex pattern that demanded understanding. Once mastered, this knowledge of the sky provided a sense of control and predictability over the physical and supernatural realms.

The sky vision shaped many aspects of their lives. It inspired stories of the world's creation and the emergence of humanity on earth. It produced a very different attitude to the passage of time from our own, driven as we are by precise clocks and calendars. The sun and stars offered a rhythm for the annual cycle of hunting and planting, and for the proper staging of religious ceremonies. The appearance of stars and planets suggested the existence of superhuman sky beings who were often associated with the power of chiefs and priests on earth, justifying the status and authority of such people. In the sky vision—the distinctive ordering of the cosmos formed by each ancient culture—human affairs were perceived as inextricably interwoven with the forces of nature. Although astrology is still popular today, it is a pale remnant of this outlook, and we can hardly realize the all-embracing power of the sky vision over the lives of prehistoric people.

The study of ancient astronomy is therefore not limited to the purely technical matter of how prehistoric observations were carried out. The most interesting question is how the skywatchers connected their skills with both everyday needs and spiritual impulses. By trying to reconstruct their sky visions, we gain an insight into how people in the distant past regarded themselves as well as the natural surroundings in which they found themselves.

The ideas of the Babylonian astronomers are particularly important because their highly precise predictions lie at the root of Western science. Vital fragments of their knowledge eventually passed into the libraries of the classical Greek philosophers and so survived to inspire some of the earliest great figures of our own scientific tradition, such as Hipparchus and Ptolemy.

But understanding the origins of our modern intellectual heritage is not the sole value of studying ancient astronomy. Besides the Babylonians, there were countless other stargazers whose efforts played no part in the growth of Western science. The idea of highly precise or systematic observations was not always the central concern of these skywatchers. Although they were just as capable of logical thoughts and arguments as we are, their sky visions did not emphasize a preoccupation with accuracy and technique so typical of present-day science. What, then, were their reasons for studying the heavens?

The first of these vanished traditions of astronomy to attract the attention of modern investigators is represented by the standing stones and stone circles of prehistoric Europe, dating back some four thousand years or more. Beginning with the discoveries of Gerald S. Hawkins at Stonehenge in 1963, it became clear that some of the stones were lined up to indicate risings and settings of the sun and moon. In the first part of this book, I present the fragmentary evidence of the ritual purposes apparently served by these ancient stones. It is also my aim to show how much confusion has resulted from recent attempts to treat prehistoric skywatchers as if they were methodical modern astronomers.

During the 1970s, the European investigations sparked renewed interest in the traditions of the New World, which offer even more penetrating insights into the nature of ancient astronomy. In the Americas, we have not only the ruins of structures, but also the testimony of native writings and inscriptions, the comments of sixteenth-century Spanish conquerors, and modern accounts of such peoples as the Hopi and the contemporary Maya, who still follow long-established beliefs and rituals. This combination of different sources is particularly striking in the North American Southwest, where ceremonial dances still continue in the villages, or pueblos, and where ancient ruins form stark, dramatic landmarks in the deserts of Arizona and New Mexico. I have concentrated on the Southwest to illustrate the depth and complexity of skywatching lore among preliterate peoples—a fact that is frequently overlooked because of all the attention paid to the achievements of literate astronomers in civilizations such as those of the Babylonians, the Chinese, or the ancient Maya.

Stonehenge is the most widely known symbol of prehistoric astronomy. But how deep were its builders' knowledge and awareness of the sky?

The Pueblo people of the North American Southwest continue to stage impressive ceremonies in which their dancers impersonate Katchinas, or ancestral rain spirits. Adam Clark Vroman photographed this line of Katchina dancers in the pueblo of Zuni, New Mexico, in 1899. Such photography is strictly banned by the pueblo dwellers today.

In the third part of the book, I will discuss the Maya in detail as an example of the achievements that were possible once a system of recording numbers and dates had been invented. What conditions were necessary for the growth of their high level of expertise in predicting eclipses and planetary motions? It is clear that no single force propelled the Mayan intellect (certainly not a recognizable scientific urge). They were motivated instead by a complex series of interwoven beliefs about the fate of supernatural Mayan ancestors and the fortunes of their living rulers. The recognition of how closely Mayan astronomy was involved with political forces has been one of the most exciting recent developments in the study of ancient astronomy.

All the skywatching peoples discussed in this book had one common characteristic, whether their expertise found its expression in great monuments like Stonehenge or the ziggurats, or was simply committed to memory, as in the case of the sun priests of the North American Southwest. They all shared a unifying vision in which their sky observations helped to order and inspire ceremonies, social groupings, food-gathering activities, and many other aspects of their lives. In the final part of this book, I will compare the examples of skywatchers in both the Old and New Worlds in order to show some common features in the outlook of these peoples.

The study of ancient astronomy is more than just an entertaining first chapter in the history of science. It offers a unique glimpse into human lives and thought processes before the rise of modern civilization.

An elaborate panel of hieroglyphs carved in stone, from the ancient Mayan site of Palenque in Chiapas, Mexico. Much of the site dates to the 7th-8th centuries A.D.

The ancient Babylonians built mud-brick stepped towers, "ziggurats," that symbolized the connection between mankind and the gods. The ziggurats inspired the Biblical story of the Tower of Babel, here depicted in a painting by Pieter Brueghel the Elder dating to about 1568.

PART I

Introduction

1. Ancient Astronomy and the Roots of Science

OUR DEBT TO THE BABYLONIAN ASTRONOMERS

The Rosetta Stone, now displayed at the British Museum.

A time framework is essential to our daily thoughts and actions. The origins of our present system are ancient, with roots in several different early civilizations. Whenever we look at a clock, we refer to units of time originating in the Babylonian number system based on sixty. To the ancient Egyptians, we owe the invention of a twenty-four-hour day and a 365-day year. The division of the calendar into its present pattern of weeks and months arises from the efforts of Babylonians, Egyptians, Romans, and Hebrews to regulate their affairs. The twelve signs of the zodiac are essentially the same ones invented by the Babylonians over twenty-five hundred years ago. All these contributions of ancient skywatchers to the modern world are so much a part of everyone's experience that we take them for granted.

How was it possible for the early astronomers to develop such enduring systems of measuring time? The answer lies partly in the traditions of writing and recording practiced by the scribes of Mesopotamia and Egypt; in the case of both civilizations, literacy had been achieved well before 3000 B.C. Until the last century the secondhand reports of classical Greek and Roman authors were the only sources of information about these traditions.

The first great breakthrough in deciphering an ancient script was the interpretation of the Rosetta Stone, which had been discovered by French officers during Napoleon's campaign in Egypt in 1799. After the surrender of the French, the Rosetta Stone fell into English hands and eventually reached the British Museum. The smooth black slab of basalt bore three inscriptions, one above the other, each one identical except that it was composed in a different script. One inscription was in Greek, and it led scholars to make the first faltering steps in translating the Egyptian demotic and hieroglyphic texts also present on the stone.

By the middle of the nineteenth century, interpretations of the hieroglyphs had reached the point where it was possible to compile a chronology of events in ancient Egypt, based on inscriptions copied from a variety of monuments. The German scholar and adventurer responsible for this attempt to reconstruct history, Karl Richard Lepsius, was also the first to recognize astronomical references in many of the engravings on tombs and temples. During the next two decades, another German Egyptologist, Heinrich Brusch, assembled enough astronomical inscriptions to fill an entire volume. This collection of texts was published in Leipzig in 1883 and was for many years considered the essential source of information about Egyptian astronomy.

At about the same time, an equally momentous advance was made in the recovery of Babylonian astronomy. During the 1870s and 1880s, a Jesuit father named Johann Strassmaier was working patiently in the British Museum, accumulating notes on several hundred inscribed clay tablets—a mere fraction of the tens of thousands of such tablets arriving at the museum from the Near East. Most of them had been discovered casually by laborers extracting mud bricks from ancient sites for construction purposes, so that few clues to their exact source were available. In fact, the museum officials purchased the majority of tablets from antiquities merchants in Baghdad. By the time they arrived in Britain, many had been reduced to clay fragments bearing the strange spidery characters of the cuneiform script. Babylonian and Assyrian scribes had created these wedge-shaped symbols by pressing the sharp end of a reed or wooden stylus into the damp clay.

By 1860 cuneiform writing had been deciphered, and the three ancient languages expressed by the script could be interpreted. There was no single key like the Rosetta Stone available to "crack" the script. Instead, a group of scholars—notably, Henry Rawlinson in Baghdad, Edward Hincks in Dublin, and Jules Oppert in Paris—persevered for many years with laborious study and inspired guesswork.

It was not until 1880 that scholars suspected the existence of astronomical texts, thanks to Strassmaier's painstaking scrutiny of the British Museum tablets. To confirm his suspicions, Strassmaier wrote to another Jesuit, Joseph Epping, who was teaching mathematics in Quito, Ecuador, at the time. Epping was at first reluctant to assist his colleague in London. After only a single year of collaboration, however, Strassmaier and Epping managed to demonstrate that Babylonian scribes had predicted the motion of the moon with startling mathematical precision. The two scholars identified the names of the major planets and of the constellations of the zodiac, so paving the way for future discoveries.

Since these pioneering efforts during the last century, only about seventeen hundred Babylonian astronomical tablets have been published; many thousands more still lie unrecorded in the storage rooms of museums throughout the world. Furthermore, the ones that have been studied belong mainly to the last few centuries of observations (from about 400 B.C. onwards), by which time most of the advanced astronomical techniques had already been developed.

Although these tablets represent the actual documents used by the Babylonian scribes to calculate the timing of eclipses and the movements of the moon and planets, their value to the historian is quite limited. We know little of the theories, motives, and interpretations of the astronomers, or how their ideas may have changed over the centuries.

It is certain, however, that Babylonian skywatching was an elite activity. Only intensively trained scribes could master the five hundred basic signs of cuneiform writing; the priests and kings were illiterate, relying on the scribes to inform them of portents and predictions related to the sky. The scribes were secular officials, but their quarters were often attached to a temple. Later on, intellectual skills seem to have been passed down among a few aristocratic families. The exclusiveness of their knowledge ensured a remarkable continuity in the keeping of records. When the Assyrian scholar-tyrant Ashurbanipal (668–630 B.C.) founded his great library at Nineveh, containing tens of thousands of tablets, he took care to preserve astronomical documents from the Old Babylonian Period more than a thousand years earlier.

It is from this period—the age of the famous lawgiver Hammurabi and his dynasty—that the first evidence of regular observations has survived. During the reign of King Ammisaduqa (1646–1626 B.C.), the dates that Venus became visible or disappeared as the Morning and Evening Star were noted in an orderly fashion, together with a list of omens associated with the planet's movements. Thereafter, a gap of nearly one thousand years exists before the next evidence of systematic observations. In the reign of King Nabonassar, around 750 B.C., the scribes began to keep a continuous record of eclipses—or so the Greek astronomer Ptolemy of Alexandria claimed in his influential treatise known as the *Almagest*, written in the early second century A.D. Ptolemy's account is confirmed by actual surviving fragments of eclipse tablets that are thought to date roughly to this period.

For another seven hundred years, until at least 50 B.C., the scribes of Babylon kept their astronomical "diaries." On these tablets the notable night-sky events of the month were reported

A Babylonian clay tablet recording a continuous sequence of observations of the planet Mercury from about 389 to 374 B.C. Similar texts exist for Venus, Mars, Jupiter and lunar eclipses.

alongside weather phenomena, commercial rates for common items in the market (such as barley, dates, and wool), changes in the level of the river, and current happenings of local or national interest. Skywatching was clearly related to the everyday world of economic and political affairs, as well as to the actions of the gods.

So when a true system of mathematical astronomy at last emerged in Babylon around 400–300 B.C., the scribes were able to draw on centuries of sustained records and accumulated experience. Nevertheless, the secret of their work lay in ingenious arithmetic rather than in extraordinary powers of observation. Although eclipses were timed with water clocks and described in detail, other phenomena were not always reported so reliably. Indeed, Ptolemy himself complained about the poor quality of Babylonian planetary data. While many modern travelers in Iraq have commented on the brilliance of the night sky, the desert horizons are in fact often obscured by haze and dust storms, and this was probably responsible for many of the astronomical errors. It was only because the scribes had a long series of observations at their disposal that they were able to reduce the effect of individual errors to a negligible level, and so arrive at their remarkably precise results.

To put it simply, their method was to consult the lists of eclipses, lunar phases, and planetary movements compiled over many centuries, and to search for regular patterns among the timings of these events. The number patterns that resulted from the search were then projected into the future. The predictions were drawn up on clay tablets in a form identical to that of a present-day astronomical or nautical almanac (usually called an "ephemeris").

As far as we know, there was no attempt to visualize the sky events in terms of orbits or any other geometrical concepts. In other words, the scribes were apparently unconcerned with explaining *why* an astronomical phenomenon took place, but their ingenious play of numbers ensured that it would not take them by surprise. Their achievement was a triumph of logical computations, not inquiring ideas.

SOOTHSAYERS AND SKEPTICS

The Babylonian reliance on numerical methods is understandable, considering that they had practiced sophisticated arithmetic as far back as 1800 B.C. At this early stage, there already existed tables for multiplication, division, squares, square roots, cubes, reciprocals, exponential functions, and many other mathematical procedures. The scribes even used algebra to tackle problems familiar to us from geometry (such as Pythagoras' theorem, which they knew a full thousand years before Pythagoras). The manipulation of numbers came naturally to them.

What, then, drove the scribes to apply their mathematical skills

The clay tablets of the Old Babylonian Period (from about 1800 B.C.) include elaborate mathematical problems, often connected with engineering and surveying tasks. In this detail from a tablet in the British Museum, a drawing of concentric circles depicts a village with a surrounding ditch and dike, the subject of a problem posed in the text.

Opposite page, bottom, Babylonian sky gods engraved on a stone cylinder seal from Tell Asmar, central Iraq (here seen rolled out as a wax impression). At the right, a worshipper brings a sacrificial kid to the sun god Shamash, while Ishtar (Venus) stands in the center holding a mace and a scimitar. Top, Assyrian seal impressions of the 9th-8th century B.C. depict worshippers presenting offerings to the gods in front of ziggurats.

with ever-increasing accuracy to the records of the sky, particularly the moon? The answer seems to lie partly in the bureaucratic demands of the Babylonian state, which must have included an efficient framework for the running of political and economic affairs.

The task of creating a well-organized calendar preoccupied the skywatchers of many early civilizations, and it was by no means a simple problem. The Babylonians, like many other ancient peoples, followed the waxing and waning of the moon, the most obvious timekeeper in the sky. Their calendar was based on the moon's phases, so that the evening on which the thin lunar crescent first appeared also marked the start of the month. But the rhythm of the moon does not perfectly match that of the sun. The interval from one new moon to the next is not always the same; sometimes it will be twenty-nine days, on another occasion thirty. The Babylonians needed to predict this interval accurately so that they could determine the evening on which their next month was due to begin.

Anticipating when the first lunar crescent would appear was a complex and demanding problem. It must have encouraged the scribes to acquire precise knowledge of the positions and speeds of the sun and moon as they traveled through the sky. The need for an orderly calendar must have been one of the motivations behind the elaborate computations of the astronomers.

The same detailed knowledge of the sun and moon was also useful for predicting eclipses. But in this case the observers cannot have been concerned with the calendar, since eclipses do not occur in a straightforward annual pattern. Instead, the eclipse calculations were probably connected with the Babylonians' interest in omens and prophecies. Although the eclipse tablets contain no actual references to astrology, the ability to know when an eclipse was due would surely have been a matter of enormous concern to the hierarchy of priests and rulers. Perhaps there was a demand for increasingly detailed forecasts. At any rate, by 200 B.C. the scribes could present remarkably precise information to the authorities. They not only knew the exact day of a lunar eclipse, but could also judge whether it was to happen during the day or night, and estimate roughly how long it would last.

Was astrology, therefore, the main reason that the skills of the Babylonian skywatchers rose to such heights? There is no doubt of the Babylonians' constant concern with portents at every stage in their history, including unnatural births and freak storms as well as events in the sky. Yet there was no actual system of astrology, not even horoscopes for individuals, until about 400 B.C. By that time elaborate computations were also becoming important in the astronomers' predictions. In fact, astronomy and astrology seem to have developed side by side as inter-related, complementary skills. The increasing use of mathematical techniques affected both activities.

By the beginning of the Christian era, a distinction seems to have crept in between astronomers and astrologers, to judge from classical accounts like that of the Greek historian and geographer Strabo: "In Babylon, a settlement is set apart for the local astronomers, the Chaldaeans, as they are called, who are concerned mostly with astronomy; but some of them who are not approved of by the others, profess to be astrologers."[2] Casting fortunes was evidently not the sole aim of Babylonian astronomy. It appears that the scribes at this late date had acquired a desire for abstract understanding and had come to pursue astronomy as knowledge for its own sake. At any rate, none of their theories or explanations for the celestial events that gave rise to their calculations have survived, if indeed any existed.

Meanwhile the conquests of Alexander the Great had exposed the Greek world to the influences of Mesopotamia and Egypt, and so, from the late fourth century B.C. onwards, Greek philosophers came under the impact of Babylonian learning. The Greeks had no long-standing tradition of recorded observations. Like the Babylonians, however, they did indulge in magical astrological practices as well as sober astronomical predictions.

In fact, the most influential astronomer of the ancient world, Ptolemy of Alexandria, was himself an astrologer. After compiling the *Almagest* around 140 A.D., Ptolemy then wrote a four-volume astrological treatise known as the *Tetrabiblos*. In his

The work of Ptolemy of Alexandria profoundly influenced European thought in the early Renaissance. This woodcut depicts Ptolemy showing an astronomical globe to his handmaidens. It appeared in 1482 as the frontispiece to Poeticon Astronomicon, *the first printed book on astronomy.*

introduction to this work, Ptolemy carefully distinguishes between forecasts related only to the sky and those also connected
to the human world. The task of foreseeing celestial motions is
more sure and effective, yet Ptolemy states that it is also quite
possible to read human fate in the stars. In this sense, the Greeks
were no more "rational" than the omen-seeking Babylonians or
the newspaper reader of today who places faith in the correctness
of the daily horoscope.

Nevertheless a fundamental difference *did* exist between the
Babylonian and Greek traditions. At an early date, from at least
the sixth century B.C. onwards, it was possible for Greek writers
to challenge popularly accepted beliefs and practices, notably in
the field of medicine, so that the term "magic" came to be used in
a disparaging way to refer to quacks and soothsayers.

So far as we know, this element of skepticism and open debate
had no place in the work of the Babylonian astronomers. Only
the Greeks developed an interest in *why* the planets followed
their wandering paths through the heavens, as well as a simple
concern with foreseeing such movements accurately. They had
clearly progressed beyond the stage of viewing events in the sky
merely as the result of the impenetrable actions of the gods. Now
there was a notion that the natural order represented some*thing*:
an impersonal reality that could be investigated, tested, and
explained. Mathematics could be used not only for the purposes
of prediction, but also to demonstrate whether or not a particular
idea was correct. Of course, Greek philosophy was often riddled
with fantastic conjectures and sloppy arguments, yet the concept
of theory and proof was a major contribution to human thought.
Here was the foundation of the methods essential to scientists
today.

Why did speculation about the universe begin in Greece, rather
than in Babylon, Egypt, or anywhere else in the ancient world?
Greece was not a single monolithic state, but a multitude of
separate, self-governing cities. From about the seventh century
B.C., the inhabitants of each city engaged in debates on the
constitutional and legal systems they were to adopt. The ordinary
citizen participated fervently in these arguments, quite unlike the
autocratic atmosphere in which the Babylonian lawgiver Hammurabi had imposed his codes. Politicians grew accustomed to
challenging and defending their views in public, and it was not
surprising that Greek philosophers made frequent comparisons to
politics and the law when they discussed medicine and astronomy. The struggle to find a well-ordered system of government
must have suggested that there were also "rules" about the
universe to be discovered.

So the emergence of democracy made the scientific attitude
possible. The skywatchers of all the other great civilizations—the
Maya and the Chinese as well as the Babylonians and the
Egyptians—labored under a rigid, unwavering framework of
earthly and divine authority. Under these enduring hierarchies,

the astronomers were able to keep consistent records for century after century, but any urge to inquire and explain was paralyzed.

In ancient China, for example, nearly all those who engaged in astronomical work were in the service of the state. They were organized in a government department which, for some two thousand years, presented regular reports to the Imperial Court on celestial events and omens. This bureaucratic system led to the careful recording of unusual phenomena at a remarkably early date; comets, for example, were noted from 613 B.C. onwards, and sunspots from 28 B.C. For all their attentiveness to the sky, however, the Chinese offered no more than the vaguest suggestions about the mechanics of the solar, lunar, and planetary movements. Like the Babylonians and the Maya, they never acquired a system of geometry, and so were unable to investigate the appearances of changes they observed in the sky. Their speculations were stifled under the dead weight of bureaucracy and mythology.

Opposite, a remarkable astronomical device dating to 87 B.C., recovered by sponge divers off the island of Antikythera near Crete in 1900. The heavily corroded gear wheels were studied by Derek de Solla Price of Yale University, who showed that they corresponded to solar and lunar cycles.

The legendary Hsi and Hso brothers, the first Chinese astronomers, receive their commission from the Emperor Yao to organize the calendar and observe the heavens. From its beginnings, Chinese astronomy was linked to the bureaucratic apparatus of the state.

The Greek development of geometry freed astronomers from their narrow preoccupation with numbers, and projected their thinking into three-dimensional space. From the start, Greek philosophers tried to envisage "working models" of the sky to reproduce the movements of the sun, moon, and planets. Their early attempts often involved much ingenuity, such as the universe of twenty-seven spheres spinning around the earth proposed by Eudoxus, but they had no way of proving that such models were correct.

The crucial step beyond fantasy and speculation was not taken until about 150 B.C., when Hipparchus—an astronomer from Rhodes whose original works survived mainly in the comments of Ptolemy three centuries later—created a plausible, detailed simulation of what he observed in the sky. Furthermore, he went on to *prove* the correctness of his model by applying the calculations of trigonometry (a branch of mathematics which he probably invented especially for this purpose). It was left to Ptolemy to improve and extend this work, but Hipparchus had achieved the vital breakthrough: the first testable astronomical theory.

The Greek invention of our scientific method is so significant that we tend to overlook how much the Babylonians had contributed to it. Although Hipparchus made careful firsthand observations of the sky, he based his calculations on figures extracted from Babylonian astronomical texts to which he had somehow gained access. It is also likely that he borrowed from the Babylonians the basic idea of applying numbers to predict the motions of the sun and moon. The Babylonians apparently had no concern to test, prove, or explain their work as the Greeks did. Yet the systematic way in which they used mathematics to project their observations into the future was the true beginning of our scientific tradition.

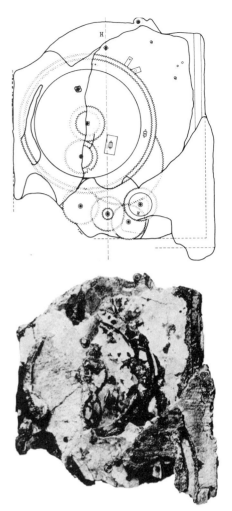

In our everyday lives we still encounter the influence of the Babylonians, for there are sixty minutes in an hour and 360 degrees in a circle. The later Greek astronomers adopted for their calculations the Babylonian number system based on sixty, and so ensured its survival to the present day.

Our calendar, however, we owe to the example of the Egyptians, not the Babylonians. For all their ingenuity in forecasting the arrival of the crescent moon, the Babylonians had a clumsy and complicated calendar, since it required the insertion of an extra month every two or three years. This was also the rule followed by the Greeks, although there the result was even more impractical, because the particular system adopted in each city depended on the whim of local politicians.

What was needed was a framework that did not rely on the awkward rhythm of the waxing and waning moon. A smoothly functioning calendar required that one conceive of the year as a fixed length of time, divided into artificial months of equal length

TIME AND THE PYRAMID: THE LEGACY OF EGYPT

regardless of the actual appearance of the moon. This method would eliminate the confusing insertions of extra months at irregular intervals, and the entire sequence of months would keep pace with the passage of the seasons.

The Egyptians hit upon such a solution as far back as the third millennium B.C. In their solar, or "Civil" year, they ignored the phases of the moon and instead devised twelve neatly ordered months of exactly thirty days each. They added an extra five days at the end of the year to bring the total up to 365. Although the Egyptians persisted in following more complicated lunar calendars for the timing of their religious ceremonies, the simplicity of the Civil Year made it ideal for the administration of their everyday affairs.

In fact, the 365-day year was so efficient and widely admired that when Julius Caesar decided to reform the calendar of the Roman Empire in 46 B.C., he employed an Egyptian astronomer, Sosigenes of Alexandria, as his adviser. The result was a calendar that closely followed the Egyptian Civil Year. Caesar arranged the "leap year" correction in February familiar to us, bringing his new calendar more exactly in line with the true length of the year. In its essentials, this calendar is the one we still keep today.

The second legacy of the Egyptians is the twenty-four-hour day. Their observers waited for the rising of particular stars to mark the passage of time. There were thirty-six of these special timekeeping stars, known as decans, used to subdivide the calendar into ten-day periods. The observers would watch for the first brief appearance of a decan star at dawn before it faded from view in the brightening sky; the first morning of the year on which this happened signaled the start of another ten-day period.

The same sequence of stars was also used to tell the time at night. As each decan star rose above the horizon, it indicated the beginning of the next "hour." Since there was a total of thirty-six decans, we would expect there to be an eighteen-hour day and an eighteen-hour night, but in practice the number of decans visible during the night would vary according to the season. At midsummer, coinciding with the arrival of Sirius, the brightest star in the sky, and the all-important annual flooding of the Nile, the night was only twelve decans long. When the Egyptian astronomers eventually decided to fix a standard length for the hour, they based their calculations on this critical midsummer timing, and so both the day and the night came to be divided equally in twelve parts.

Besides these advances in practical time measurement, the Egyptians also left us an enduring symbol of ancient knowledge—the Great Pyramid. Like Stonehenge, it has inspired endless theories about its astronomical and mystical functions, often of a fantastic nature. Speculations began in earnest when Charles Piazzi Smyth, the Astronomer Royal of Scotland, published his book *Our Heritage in the Great Pyramid* in 1864. Among his many wild claims, Smyth proposed that the dimensions of the pyramid

Opposite, the Egyptian sky goddess, Nut, was envisaged as a woman bending over the heavens with hands and feet touching the horizon. She was the mother of Re, the sun, whom she swallowed in the evening and gave birth to the next morning.

*A nineteenth-century view of
the Great Pyramid.*

were chosen so that they recorded most of the laws known to
modern physicists, which the Egyptians had supposedly discovered by centuries of careful study. He also theorized that the
layout of the interior passages was based on astronomical observations, though to figure this out, he used a date for the building of
the pyramid that was at least five hundred years too late. (The
correct construction date is thought to be about 2650 B.C.). Since
the publication of Smyth's theories, countless others have advanced claims about the scientific and spiritual properties of the
pyramids.

A cool appraisal of the evidence suggests that only the most
rudimentary astronomical knowledge is incorporated in the Great
Pyramid, although it remains a staggering feat of ancient engineering. The four sides of the base, each about 756 feet long, line
up with the four cardinal directions. The greatest error in the
orientation is little more than one-twelfth of a degree, equivalent
to one-sixth of the face of the full moon as we see it from the
earth. It is astonishing that the pyramid builders managed to
maintain such accuracy as they manhandled more than two
million limestone blocks, each weighing an average of over two
tons. Yet the task of setting out the original north–south line was
an elementary exercise. There was no Pole Star shining due
north in the heavens as there is today. However, the observers
could simply mark the rising and setting points of a star in the
northern sky, then halve the angle between these points to arrive
at their north line. If this simple observation was carried out
carefully, it would have provided quite enough accuracy for the
north–south sides of the pyramid.

Many other astronomical claims have been made about the
baffling series of rooms and passages inside the Great Pyramid.
Perhaps the most convincing theories explain the two so-called
"air shafts" leading from the main chamber. These shafts slope
upwards toward the outside of the pyramid, but they could not
have been observation tubes; at both ends of each shaft, there is a

horizontal bend so that a direct sightline to the sky is impossible. Nevertheless, the main angle of each shaft does coincide with the north–south passage of two important stars at the time of the pyramid's construction—Thuban, the star then closest to the North Pole, and Alnilan, a star in Orion's Belt. We know that the Egyptians identified Orion with Osiris, the underworld god of vegetation and rebirth. Temple inscriptions speak of the pharaoh's ascension to join Osiris in the sky, where he was to command the eternal revolutions of the stars. If this explanation is correct, then the two shafts were not intended for practical skywatching or ventilation, but for the passage of the pharaoh's soul on its journey toward the afterlife in the heavens.

To judge from images on coffin lids and temple walls, the Egyptians' awareness of the sky was inseparable from their constant preoccupation with the cycle of death and rebirth—a complex religious outlook. Yet compared to the Babylonians, the intellectual accomplishments of their astronomers were crude and insignificant. They played no part at all in the growth of the mathematical and predictive techniques perfected by the Babylonians and Greeks. Only their sensible time-reckoning schemes survived, while the mainstream of early science passed them by.

Because of the vital contribution of the Babylonians and Greeks to our own outlook, we naturally emphasize their achievements. But their systematic attitude to the sky was in every way exceptional. The example of the Egyptians is actually far more representative of most of the skywatching traditions presented in this book. Despite its unique scale, the Great Pyramid is typical of the structures left by the astronomers, architects, and engineers of the ancient world. It was clearly not an observatory, since one could not see through the two shafts. Instead, astronomical alignments seem to have been involved in the design of the pyramid as a way of expressing the Egyptians' beliefs.

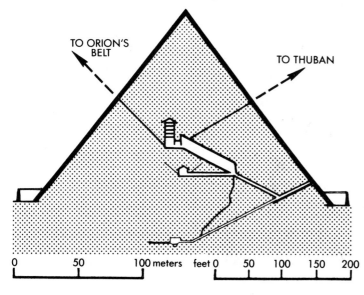

Cross section of the Great Pyramid, showing two astronomically aligned air shafts.

Like the builders of Stonehenge or the Mayan priests described later in this book, the Egyptian astronomers were not practicing "science" as we understand it today. They did not pursue knowledge as an end in itself, detached from spiritual values and social reality. Rather, their sky lore served to reflect and reinforce their most essential attitudes. The architects used astronomical planning as an effective way of communicating religious ideas. By constructing such impressive monuments as Stonehenge or the pyramids, ancient rulers created powerful expressions of the values that bound their societies together.

SKY VISIONS, ANCIENT AND MODERN

The astronomers of the great literate civilizations naturally dominate our attention, but there were other skywatchers in the ancient world who had no tradition of writing or recording of any kind. This unwritten prehistoric lore did not contribute to the growth of science, yet neither did its influence pass away without trace.

Many of our popular folk customs spring from a curious mixture of pagan European beliefs and practices now hidden beneath the conventions of Christianity. For example, the lighting of candles on the Christmas tree and the burning of the Yule log are reflections of the time, only a century ago, when pagan bonfires still blazed at midwinter throughout the European countryside. Our calendar also preserves the framework of the solar festivals of ancient Europe. Before the coming of the Romans, there were four such festivals, known in Irish tradition by the following names: Imbolc (February 1), Beltane (May 1), Lughnasa (August 1), and Samhain (October 31–November 1). These dates represent a division of the solar year into eight equal parts, since they fall midway between midsummer, midwinter, and the equinoxes. According to folklore, though, the exact timing of at least two of the festivals was determined by the phase of the moon.

Superstitions and folktales linger around the megalithic monuments of Europe. Romantic artists of the last century often imagined them as the haunt of gory priests practicing human sacrifices, as seen here.

We still recognize some of these celebrations in the Scottish Quarter Days and in the American observance of Ground Hog's Day and Halloween. In particular, our Halloween customs retain the original fearful associations of Samhain, when it was believed the spirits of the dead Celts walked again. Only a hundred years ago, Scottish mothers worried that on this night goblins might try to steal their babies, while in Ireland magical armies were thought to emerge from caves and ancient burial mounds to wreak havoc and destruction.

Superstitions of this kind are unconscious fragments of pre-Christian religious beliefs. Many such legends linger around the prehistoric monuments of Europe—notably, the tombs and rings of standing stones raised during the Neolithic or New Stone Age, which lasted from about 5000–2000 B.C. For example, the inhabitants of the island of Lewis in the Scottish Outer Hebrides recall many tales about the ring of stones known as Callanish, which overlooks Loch Roag on the island's bleak Atlantic coast. One legend speaks of a "Shining One" who walks down the long avenue of gaunt stones as the cuckoo's first call is heard in the spring. Is it possible that stories like this one were inspired by a genuine tradition of solar rites that took place among the stones thousands of years ago?

Archaeologists, astronomers, and amateur enthusiasts have studied the standing stones, hoping to discover traces of a prehistoric system of astronomy. These efforts began as long ago as the turn of this century. At that time, a distinguished astronomer, Sir Norman Lockyer, traveled extensively throughout the British Isles, measuring the lines marked out by prehistoric tomb passages, burial mounds, stone rows, centers of stone circles, and outlying stones. Lockyer became convinced that a single calendar scheme united the builders of all these ruins. From Cornwall to

The standing stones of Callanish, situated on the island of Lewis in the Scottish Outer Hebrides.

Aberdeenshire, Lockyer believed, prehistoric people had celebrated the same solar festivals as had eventually passed into the folklore of the Celts some two thousand years later; to commemorate these events, stones were lined up in the direction of the sun or stars as they rose or set on the appropriate day. Lockyer supposed that a class of astronomer priests had once controlled the whole of prehistoric Britain, "on whom the early peoples depended for guidance in all things, not only of economic, but of religious, medicinal and superstitious value."[3]

Since Lockyer's day, a number of investigators have advanced even more ambitious claims about the intellectual skills of the ancient Britons. On the other hand, some critics point out that "Stone Age science"—a precise, unified system of knowledge—is unlikely to have appeared in preliterate Europe. Such arguments continue to divide those interested in the standing stones.

The investigation of an astronomical structure is, in theory, a simple affair. It involves the measurement of a suspected observing line with the aid of a surveying instrument such as a transit or theodolite. The researcher can then easily find the astronomical position (or "declination") of the surveyed line by consulting a pocket calculator. It is also quite simple to allow for the slight change in the positions of the sun, moon, planets, and stars that has occurred since prehistoric times.

The problem here is not the basic technique of the investigator, but the interpretation of the results. Without written records, we have no way of knowing if a line of stones was deliberately set up by the original builders to indicate an astronomical event. The line may, for instance, point to where the sun rises on the horizon at midsummer, but how can we be sure that the prehistoric skywatcher was aware of this? If we suppose that the alignment *was* deliberate, it is still uncertain whether the original observation was carried out crudely or accurately. And, finally, there is the motive of the observer to consider; was the line of stones part of a rational program of investigation, such as a scientist would devise today? Or was it the work of a pious priest, driven to express his culture's mythological ideas about the heavens?

We can begin to answer some of these questions by considering a variety of sources. Investigations of the alignments to the sun and moon, archaeological digs at the ancient sites, and even the traditions of present-day folklore all provide clues to the purpose of the prehistoric stones of Europe. The standing stones do seem to be connected with Neolithic peoples' ideas about fertility and the seasonal round, and their attitudes to death and the afterlife. However, the precise intentions of the builders—the specific ways in which their sky observations served their religious values and everyday needs—remain elusive. Because of the absence of written records, substantial gaps in our understanding remain to challenge and frustrate the researcher.

In the New World, the situation is different. Skywatching customs linger on among traditionally minded Native Americans, particularly in parts of Arizona, New Mexico, Guatemala, and Mexico. Although Native Americans are increasingly affected by modern civilization, their spiritual practices continue, in some cases reflecting traditions that were established centuries ago. From the work of anthropologists, we can often gain a clear idea of how skywatching relates to ceremonies, crop raising, and special cults or privileged groups within societies—connections so often missing from the study of ancient cultures.

In addition to firsthand studies by present-day observers, other sources of information enrich our view of the Native Americans. The century-old reports of the first pioneering North American anthropologists are obviously valuable, for the ceremonial activities of some tribes were less disturbed in the 1880s and 1890s than they are now. And in Mexico and Peru, a handful of European commentators (usually Spanish priests) took the trouble to note their observations of indigenous customs as far back as the sixteenth and seventeenth centuries.

Perhaps the most interesting sources of all are documents and inscriptions left behind by the native priests. These records include numerous texts of the Mayan civilization, which flourished in Central America from about 300 A.D. until the arrival of the Spanish in the sixteenth century. The Mayan scribes practiced a curious form of hieroglyphic script which has only been partially deciphered. Their inscriptions are found mainly on engraved stone tablets fixed to temple walls or erected as freestanding pillars, or stelae. In addition, three or four Mayan books have survived, the pages made of bark and joined together like the folds of an accordion. These books (described more fully in Chapter 16) are intriguing because they seem to be the actual manuals or almanacs consulted by Mayan skywatchers—the equivalent, in fact, of the Babylonian clay tablets.

One of these documents lay neglected in the archives of the public library in Dresden, Germany, for over a century. By 1880, the head librarian, Ernst Förstemann, had realized its significance, and in that year he published the first of many ingenious studies. He recognized that the Dresden Codex contained tables

Detail of two supernatural figures depicted in the Dresden Codex.

for predicting the movements of the moon and the planet Venus. Like the Babylonians, the Maya never used geometry to examine or explain their work, but instead relied on arithmetical techniques to project the celestial cycles into the future. It is now clear that the Dresden tables provided a reliable "early warning system" for lunar eclipses, based on highly accurate, long-term observations of the moon. Although their eclipse predictions were not quite as specific and detailed as those of the late Babylonians, the intellectual accomplishment was just as remarkable. Unfortunately, these skills were not appreciated by the Spanish conquerors of the Maya, and so their knowledge was, for the most part, destroyed. To this day, however, the pre-Christian calendars and ancient ceremonial beliefs linger on in the remote highlands of Mexico and Guatemala (see Chapter 14).

During the 1970s and early 1980s, there has been a remarkable upsurge of interest in Native American skywatching traditions as a result of the combined efforts of archaeologists, anthropologists, and astronomers. From a unique blending of sources—native writings, present-day beliefs, anthropological reports from the nineteenth century, and recent surveys of ancient monuments—a remarkably complete picture of certain astronomical traditions has emerged.

I have chosen to emphasize the achievements of the Maya of Central America and the Pueblo and Anasazi cultures of the Southwest because it was in these areas that some of the most striking advances in new knowledge occurred. By examining this particularly rewarding research in detail, we can begin to see how astronomy played its part in the intricate ceremonial, social, and political life of these peoples.

I also draw attention to the remarkable variety of skywatching traditions that flourished in the New World before the impact of modern civilization. There was a richness and diversity of cultural activity which the stereotyped image of "The Indian" might lead us to overlook. Sophisticated astronomical ideas were developed by mobile hunting-and-gathering peoples as well as by settled crop growers. The sky observers were sometimes privileged priests who were members of a hierarchy of lords and nobles. But there were other skywatchers who belonged to relatively egalitarian tribes, where authority was based on age and respect, not class.

The reasons why class-divided societies developed are still not fully understood by anthropologists. Astronomical ideas were surely important in this process, since they helped to justify the actions of a ruler as part of the divine order of the cosmos. How were simple observations of the sky gradually transformed into myth and dogma that served to uphold the power of priests and despots? This is one of the most interesting and difficult questions faced by those concerned with ancient astronomy.

In trying to answer it, there is a risk that we will take the example of one particular Native American tribe as too literal a

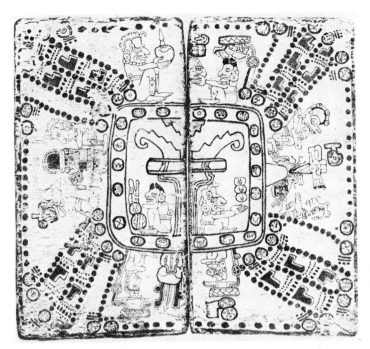

A symbolic diagram of time and space from ancient Mexico. The cross shape refers to the four directions of space, while the border of black dots refers to the time units of the sacred calendar. From the Codex Madrid, compiled in Central Mexico shortly before the Spanish conquest.

guide to the character of life in the vanished prehistoric past. Most anthropologists do not regard the indigenous peoples of today as "living fossils," as if they represented earlier stages of social evolution through which all human societies must inevitably pass. We cannot ignore the specific historical events that have shaped each culture in a distinctive way. In other words, the cultures of the Hopi and the modern Maya do not present us with "frozen images" of prehistory which we can substitute crudely for the gaps in our understanding of the ancient past.

But the example of the Native Americans does demonstrate the range and complexity of knowledge that could be acquired by a skywatcher who had little or no contact with the attitudes of the Western world. Each Native American society developed its own distinctive "sky vision," a unifying outlook which linked its ideas about astronomy to other key aspects of day-to-day life and spiritual beliefs. The self-contained spirit of modern astronomers—their urge to pursue new knowledge of the cosmos for its own sake—had no place in the scheme. Not even the ancient Maya, whose achievements represent the pinnacle of skywatching in the New World, practiced science as we understand it; on the contrary, as the ability of their astronomers to predict eclipses and planetary appearances grew more refined, so their involvement with extraordinary religious and political values intensified. This was an outlook quite unlike our own, but typical of the unfamiliar perspectives opened up by the sky visions of the ancient past.

The standing stones of Callanish, situated on the island of Lewis in the Scottish Outer Hebrides.

PART II

The Sky and the Stones

Standing stone at Barbrook, Derbyshire, in the Peak District of central England. The stone belongs to one of a number of small stone circles probably erected in about 1500 B.C.

2. Superstition or Science?

The current wave of interest in ancient astronomy began with the work of Gerald S. Hawkins, an English-born astronomer who lectured at Boston University during the 1960s. In the course of a visit to Stonehenge in 1961, he was struck with the idea that the pattern of pits, wooden posts, and upright stones might conceal an astronomical meaning. After testing this possibility with the aid of measurements and calculations, he summarized his work in two papers published in the distinguished scientific journal *Nature* during 1963 and 1964. The interest these papers generated in the scientific world was followed by a public sensation when Hawkins' popular book, *Stonehenge Decoded*, appeared in 1965.

Aside from the fascination of the topic itself, several aspects of Hawkins' work served to fuel the enthusiasm of the public and the press. One was his use of a computer to "decode" the numerous astronomical alignments he detected at the site. In fact, a computer was not absolutely necessary in performing the calculations, but this link between a modern machine and a four-thousand-year-old record of observations struck a deep chord of interest in many.

Another remarkable feature of Hawkins' work concerned the prediction of solar and lunar eclipses. Given the limited knowledge and materials at their disposal, the Stonehenge observers would have needed persistence, luck, and ingenuity to forecast the actual dates of eclipses. But Hawkins developed an ingenious theory, showing how certain features of the site could have been used to predict a cycle of eclipse danger periods. He suggested that knowing the month when the sun or moon might disappear would be a powerful instrument in the hands of priest-astronomers at Stonehenge four thousand years ago.

Hawkins' work was followed in 1967 by the publication of Alexander Thom's book *Megalithic Sites in Britain*, which pushed the claims for the abilities of the prehistoric astronomers still further. (Megalith, from the Greek, simply means "large stone.") Thom's book summarized the highly accurate surveying work he had carried out in his spare time while he was Professor of Engineering at Oxford University. His interests had been aroused during his regular sailing trips around the shores of western Scotland, where standing stones are to be seen in impressive isolation on barren crags above the lochs, or on lonely heather-covered moors.

As he drew up plans of nearly two hundred stone circles from every region of Great Britain, Thom became convinced of some remarkable theories. To begin with, he showed that while some rings were truly circular, others seemed to be based on elaborate geometrical designs, originally set out with a unit of length which he called the Megalithic Yard. It was an astonishingly precise and

MEGALITHIC MASTERMINDS

Gerald S. Hawkins at Stonehenge in 1973.

Alexander Thom, who began his pioneering investigations of megalithic astronomy and geometry nearly fifty years ago.

uniform measure; according to Thom's calculations, its value was exactly 2.722 feet (.830 meters). Indeed, there was so little deviation in this unit, Thom argued, that identical yardsticks must have been manufactured at some central "headquarters" and passed out for use all over the country. A single intelligent plan had inspired the arrangement of crude masses of stone in highly accurate geometrical patterns. This was a startling theory.

Thom's claims in the field of astronomy were even more ambitious. This tall and energetic Scotsman, now in his eighties, had for many years hiked around the most desolate regions of northern and western Scotland, shouldering heavy survey gear to measure the exact relationship of one ancient stone to another. With this equipment, Thom recorded "alignments"—that is, pairs of stones, or just one face of a stone, that seemed to point to a place on the horizon where the sun or moon rose or set. After decades of research, Thom came to believe that the prehistoric people of Britain had made truly scientific attempts to understand the motions of the sun and moon; that their irrational fears and superstitions had given way to a sustained, intellectual curiosity about the sky.

According to Thom, the purpose of the stones was not exclusively religious or ceremonial. In his recent work, many of the alignments marked by the stones are seen as an ancient "research program" dedicated to forecasting lunar eclipses. This method was a more complicated method of prediction than Hawkins', demanding much more precise observations. Nevertheless, Thom implies that one of the motives was similar: the priests wanted to impress on the ordinary population the power of their knowledge of the sun and moon. The prediction of eclipses was the most impressive undertaking of a single scientific establishment, which had apparently dominated communities from the Shetland Islands of Scotland all the way down to the Gulf of Morbihan in Brittany for century after century.

A GOLDEN AGE OR A "MENACING WORLD"?

During the 1970s, magazine articles, television films and popular books about the theories of Hawkins and Thom began to multiply. It is easy to understand their wide appeal, for the theories offer a romantic sense of continuity to the present-day reader faced by the uncertainties of our own science-dominated world. There is something reassuring in the notion that precise science was also practiced in Britain four thousand years ago; whatever the purpose of this activity, it surely cannot have been as alarming as some scientific research seems today.

Several authors believe that an exact knowledge of the sun and moon's movements was developed by ancient peoples for spiritual or magical ends far beyond our comprehension. One of the most eloquent "megalithic mystics" is the British author John Michell, who concludes that the efforts of the monument builders and the folklore associated with the stones

. . . could not just have been a product of the need for reckoning the time or the date or of some abstract desire for astronomical information. What then was the real object of Stone Age science? The answer may well be not only beyond the present imagination of the archaeologists, but incomprehensible in terms of the conceptual model of modern orthodox science, which is unlikely to have been the model referred to by professors of the magical science of antiquity.[4]

Other writers, too, suggest that prehistoric people were in touch with a power that we have lost or neglected. Janet and Colin Bord, authors of *Mysterious Britain* and *The Secret Country*, suppose that

. . . the precise calculations obtainable from the stone observatories were surely used for some purpose other than a farming calendar. It could be that the scientists of that age, by accurately plotting the positions and relationships in the heavens of the sun, moon, planets, and stars, could calculate the optimum time to capture and store the inflow of cosmic energies to be released.[5]

It is hard to deny the element of nostalgia and wishful thinking that such a picture of ancient Britain offers at the present time. If we seek a comforting explanation of the prehistoric stones, one that closely fits the pattern of our own insecurities, then the theory of ancient technicians in command of miraculous powers will do.

As my own interest deepened, I found myself distrusting such simple answers, and also wondering if the entire emphasis on "science" in studies of the megaliths might be mistaken. Could the effort to prove that out ancestors were high-precision technologists be little more than a disguised, unconscious projection

Standing stones have often inspired visions and fantasies of a prehistoric "Golden Age." William Blake believed that the Druids built Stonehenge and practiced a virtuous religion uniting all of "Albion," or Britain. His engraving of a vast imaginary megalithic archway is from Jerusalem, *published by the artist between 1804 and 1820.*

of ourselves onto the unwritten past? This suspicion also disturbed the distinguished prehistorian Stuart Piggott:

> God-like, we try to make ancient man in our own image, and the preferred image varies with the changes of taste and preference of our own society. We desire to find admired qualities in the past, and mathematical and scientific qualities are admired today. If ecstasy and shamanism were more highly regarded than these, this is what we might be looking for—and doubtless finding—in prehistory.[6]

Piggott is not the only archaeologist to react to the fashionable astronomical theories with skepticism. Many of his colleagues are accustomed to a different picture of prehistoric Britons than is implied by the researches of Hawkins and Thom. The traditional outlook of some archaeologists is shaped partly by classical authors such as Tacitus, who wrote in his *Annals* of the uncouth British priests who foretold the future by consulting the human entrails of sacrificed captives. He also vividly describes a Roman raid on Anglesey in 60 A.D. when British priests howled curses so fiercely that the troops were temporarily paralyzed with fear.

If the Britons of the first century A.D. were indeed such rough barbarians, the notion that their ancestors at Stonehenge were highly civilized mathematicians is difficult for many to swallow. The celebrated English author and excavator Jacquetta Hawkes expressed her misgivings as follows: "Could the poorly equipped inhabitants of a notoriously cloudy island have stepped, in this one direction, so far out of the cultural context in which they have always belonged?"[7] Perhaps this is merely a matter of prejudice, as biased as Tacitus' own attitude to the Britons must have been. Other prehistorians, however, have supplied good reasons for doubting the extreme claims of some of the astronomical theories.

One of them is Aubrey Burl, who lectures in archaeology at Birmingham University in central England. Burl has a profound grasp of the results of excavations undertaken at stone circles in Britain, both recently and throughout the last century. In such books as *Prehistoric Avebury* and *Rings of Stone*, this knowledge is coupled with a vivid writing style—a rare combination for a scholar. He introduces a dramatic picture of prehistoric social life which clashes with the science-dominated world implied by Thom's theories.

We are now sure that Britain's standing stones were erected over a very long period of time, perhaps as far back as the earliest farmers who were established there well before 4000 B.C. These settlers—who are generally believed to have brought their families and livestock from the Continent in open boats to the eastern and northern seaboard of Britain—did not begin to build stone circles immediately, however. Instead, the early farmers were preoccupied with religious practices that included the communal burial of bodies in elaborate stone tombs. The building of

Aubrey Burl speaking at a prehistoric site in Wiltshire in 1981.

stone rings only became widespread at a slightly later period, perhaps beginning about 3300 B.C. and continuing until at least 1500 B.C. This represents an impressive continuity of tradition, as long as that of Christianity itself. But just as Christianity has been expressed in various forms through periods of conflict and upheaval in society, so, too, the long record of the stones implies neither uniformity nor a peacefully unruffled passing of the centuries.

Burl has shown how the development of stone circles reflects a gradual change in ancient Britain from a basically communal

Map of British prehistoric sites mentioned in the text or illustrated in this book.

society to one where some individuals were more powerful than others. Most of the larger stone circles seem to date from the earlier periods. They appear to express common beliefs that bound people together on fairly equal terms, both because huge pools of labor were necessary to construct them and because the enclosed space was big enough to accommodate many people if they participated in the ceremonies. Later on, the rings became smaller. Indeed, some could have been the work of a single family, intent of creating a private worshipping place.

The meaning of the stones varied not only with the passing of time, but also from place to place. Burl's research demonstrates that regional groups of stone rings are set apart from one another by differences in situation, architecture, and the ceremonies that were held within them. The local groupings even had preferred numbers of stones. For example, Burl indicates that a circle usually had twelve stones in the rugged Lake District, while ten was the favorite number in the rolling pastures of northeast Scotland. Perhaps the most striking contrast of all is between the British Isles and Brittany. Thousands of prehistoric Breton stones were erected either as solitary monoliths or in formations such as rows and avenues, yet stone circles with architecture and finds similar to the British ones are almost unknown.

These comparisons flatly contradict the findings of Thom. As the result of his fieldwork, he insists on the presence of a single unwavering unit of length and one astronomical system, persisting for generations all the way from the Outer Hebrides to Finistère in Brittany. In contrast, archaeological evidence suggests some shared religious beliefs and the common use of stone as a building material, but widely differing regional preferences that changed over the centuries. Here lies the basic conflict between the archaeological and the astronomical theories.

In his books, Aubrey Burl expresses a personal vision of prehistoric Britain that is very different from the precisely ordered scientific utopia imagined by some enthusiasts of Thom's theories. The stone circles, in Burl's view, were not calculating machines but places where prehistoric people communicated with the unknown. Burl conducted several excavations at stone rings in northeast Scotland and discovered offerings at these sites consisting of odd little deposits of burned human bones, ashes, with a piece of pot or flint thrown in—signalling the existence of beliefs and rituals that we can barely understand, "superficially meaningless practices which may reveal the attempts of the people to obtain safety within their living and menacing world."[8]

According to Burl, there was no Golden Age in the Stone Age, and it was to religious officials such as priest-shamans or witch doctors—not to enlightened astronomers—that the people of these early farming communities turned for relief from the many anxieties and dangers facing them.

The skeletons of prehistoric people, so frequently marked by disease, do confirm something of Burl's bleak outlook on the times. Arthritis and mouth infections were particularly widespread, while Burl quotes a "dreadful list" of identified ancient British diseases, including polio, sinusitis, tetanus, tuberculosis, and probably plague and malaria as well. Those lucky enough to survive beyond their thirties were clearly exceptional, while it is estimated that four in ten died before they were twenty. Occasional discoveries of arrowheads lodged in ribs or backbones suggest that violence, perhaps even warfare, was not unknown.

From this evidence of a precarious existence, one can understand why some archaeologists regard the standing stones as a product of primitive fears and irrational religious beliefs rather than a sophisticated scientific philosophy. In *Prehistoric Avebury*, Burl expresses his view in forthright terms:

> There is not one aspect of Neolithic and Bronze Age life among the thousands of artifacts, monuments, settlements, the economies, burial practices, rituals, even the personal ornaments of the dispersed communities in Britain to suggest that there was an island-wide culture banding all these societies together, with a national yardstick and an all-powerful priesthood dedicated to a study of the heavens.[9]

Excavations at two small stone circles at Fortingall, Perthshire, Scotland, in 1970. There were prehistoric "offerings"—small amounts of charcoal and cremated human bone—deposited at both sites. A century ago, farmers attempted to demolish the stone circles; a Victorian beer bottle was found under one disturbed stone.

THE GEOMETRY OF STONE CIRCLES

The results of recent research indicate that a compromise may be emerging between the theories supporting religion and those supporting astronomy. For instance, some archaeologists (including Burl) find no difficulty in believing that astronomical observations of a simple kind *were* carried out as one aspect of the symbolic, ritual practices of the stone circle builders. Meanwhile, a new generation of scholars specifically interested in Thom's work (notably Clive Ruggles of Leicester University) have begun a thorough firsthand reexamination of the evidence at numerous standing stone sites. Their initial results confirm some of the basic discoveries made by Thom, but discredit many of his more extreme claims, particularly those involving accurate, scientific astronomical alignments.

We can appreciate the current state of research by considering briefly the problem of the geometrical layout of megalithic sites. Several statisticians agree that a unit of length does seem to exist in the measurements of diameters which Thom made in the course of his surveys of stone circles. However, the evidence comes mainly from Scottish sites, rather than the widespread distribution throughout Britain originally claimed by Thom. Furthermore, there is no compelling evidence that this unit was really as accurate as Thom supposes his Megalithic Yard of 2.722 feet to have been. In fact, the measurements of the stone circles vary sufficiently to leave open the possibility that a human body measurement, such as a pace or a person's height, may have been used.

A similar compromise verdict applies to the geometrical shapes marked by the positions of the stones. Thom's remarkable discov-

The flattened outline of the stone "circle" at Black Marsh, Shropshire, is obvious from this survey by Alexander Thom, below, left. He has superimposed an elaborate geometry of intersecting arcs. By contrast, Ian Angell's drawing, right, shows how the same shape could be produced more simply with the aid of a loop of rope pulled tightly around three pegs (marked by crosses).

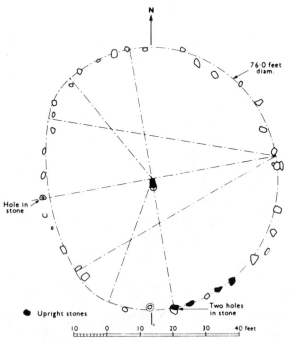

N

76·0 feet diam.

Hole in stone

Upright stones

Two holes in stone

10 0 10 20 30 40 feet

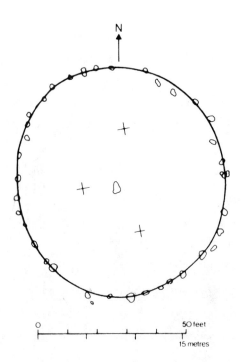

N

0

50 feet

15 metres

ery that many stone circles were deliberately set out in regular, noncircular patterns is now widely accepted, but the particular set of designs adopted by him has been seriously questioned.

Thom supposes that his megalithic mathematicians acquired a refined abstract understanding of geometry, including a knowledge of Pythagorean triangles at least two thousand years before the age of Pythagoras. This knowledge then led to their interest in recording certain elaborate patterns in the layout of the stone rings. However, Thom's theory does not prove that his constructions fit the surveys of the rings any better than other simpler suggestions. This point has been demonstrated clearly by Ian Angell of London University, who has simulated the megalithic designs with the aid of computer-drawn graphics. Angell shows how the monuments could have been set out with fairly crude arrangements of ropes and stakes in order to arrive at a pleasing shape, and with little necessity for abstract geometrical thinking.

All these revisions to Thom's geometrical theories support the notion that a certain amount of deliberate planning *was* present in the minds of the megalith builders. However, their practices of laying out distorted patterns on the ground were not so highly precise or rigidly followed as to conflict with the archaeological evidence, which points to much variety in the customs of the stone circle builders from one region to the next.

In other words, prehistoric people were not preoccupied solely with accurate standards and the pursuit of mathematical knowledge. Their incorporation of unusual shapes in the design of the stone monuments was perhaps a foreshadowing of scientific activity, which could eventually have led them to develop a genuine system of geometry. So far, though, no convincing evidence is available to suggest that they actually did achieve such a system, or that their efforts were carried out in a truly scientific way.

The task of assessing the astronomical theories is even more difficult. Although the movements of the sun and moon have not changed significantly in four thousand years, the interpretation of the evidence on the ground presents many problems. A single alignment can be dressed up as the achievement of a Stone Age genius, or disparaged as only what a moderately observant farmer might have known about the sky. It can also be dismissed as pure chance, an astronomical accident never intended at all by the original erectors of the stones.

3. Stonehenge Reconsidered

The best-known of all ancient astronomical sites is, of course, Stonehenge—a unique monument that has challenged and inspired the work of scholars, enthusiasts, and artists for centuries. Despite its fame and the efforts of so many experts to understand it, there is surprisingly little agreement about the skywatching that was practiced there. It can scarcely be doubted that astronomical activity *did* take place at Stonehenge, yet it has proved difficult to establish exactly how skillful the original observers were.

We can appreciate this difficulty by considering the most celebrated of all alignments in prehistoric astronomy. Virtually everyone who visits Stonehenge is aware that the monument faces the place on the horizon where the run rises on midsummer morning. If one stands near the center of the ring and looks toward the northeast, a battered leaning pillar of rock appears in the distance, framed by three of the great stone archways of the circle. This is the Heel Stone. (There are several different explanations of the name; according to one curious folktale, a tussle between the Devil and a friar ended with the stone striking the friar. As the stone hit the friar's upraised heel, the sun rose and the Devil

The most celebrated ancient astronomical alignment. It runs from the center of Stonehenge along its axis toward the distant Heel Stone and the midsummer sunrise.

was forced to flee.) The tip of the Heel Stone about 250 feet away nearly coincides with the bare, level expanse of the horizon. This spot is where the sun first glimmers on midsummer day. The association of Stonehenge with the midsummer sunrise was recognized long ago; in an account published in 1740, the celebrated antiquarian William Stukeley first noted that the entrance faced northeast, "whereabouts the sun rises, when the days are longest."[10]

Stukeley supposed the building of Stonehenge to be the work of the ancient Druids, the priests of the native Britons and Gauls described by Caesar and other classical writers. The idea was publicized by Stukeley's quaintly illustrated books and is still popular today, even though modern archaeological dating methods show that the main temple at Stonehenge was erected some two thousand years before the classical accounts of the Druids.

Today, on midsummer morning, ceremonies are performed inside the monument by the United Ancient Order of Druids, a religious order founded in 1833. Its members, the Druid "priests," wear white robes and indulge in solemn processions that weave in and out of the stones, while a barbed-wire fence excludes the curious and often derisive spectators. Although they have no claim to authenticity, the white robes and chants of the modern-day Druids have helped to fix in our mind's eye an image of high religious solemnity attending the astronomical event. Perhaps it was by means of a similar dramatic rite that the original priests celebrated the midpoint of the year and impressed their power upon the prehistoric farmers of Salisbury Plain. If we are inclined to laugh at the bogus ceremonies of the modern Druids, they nevertheless suggest to us how impressive a religious gathering held in the light of the rising sun might once have been.

But could the entire idea of the midsummer rituals and the sunrise alignment be a mistake, a romantic delusion? The Heel Stone, it seems, was positioned appreciably "off target." Today the sun does not rise directly in its path but slightly to the left, or north, of it. Moreover, because of an extremely slow shift in the earth's axis over the centuries, the sun would have risen *even farther* to the left back in prehistoric times. A few minutes after the first flash of light, the sun would indeed have appeared right over the top of the Heel Stone, but by that time it would also have cleared the horizon by a distance nearly equal to its own diameter. This apparent failure of the Stonehenge builders to mark the precise sunrise point is crucial to an understanding of the monument. It demands an explanation.

The task of weighing one theory against another is complicated by our ignorance of what lies beneath the turf at Stonehenge. About half the monument remains unexcavated, while much of the other half was dug up during the 1920s by the archaeologist Colonel William Hawley, whose techniques left much to be desired. When parts of the site were excavated systematically by a

A view along the axis of Stonehenge looking toward the center of the monument, from William Stukeley's Stonehenge, A Temple Restor'd to the British Druids, *published in 1740.*

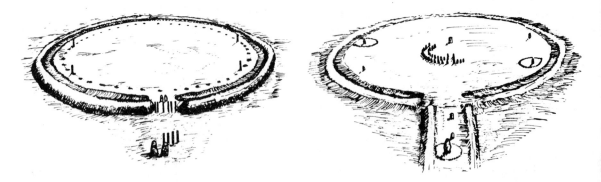

Stonehenge passed through many different construction phases. The site began around 3100 B.C., top left, with a circular ditch and bank, a ring of 56 pits (the Aubrey Holes), numerous posts erected near the entranceway, the Heel Stone and possibly its companion stone found in 1979. The entrance orientation may have been chosen because of lunar alignments. Around 2150 B.C., two concentric circles of bluestones, top right, were erected in the center of the monument, while its axis was altered to the midsummer sunrise and emphasized by an avenue of parallel banks and ditches.

Plan of the excavations directed by Michael Pitts in 1979. His digging team discovered a substantial stone-hole close to the Heel Stone.

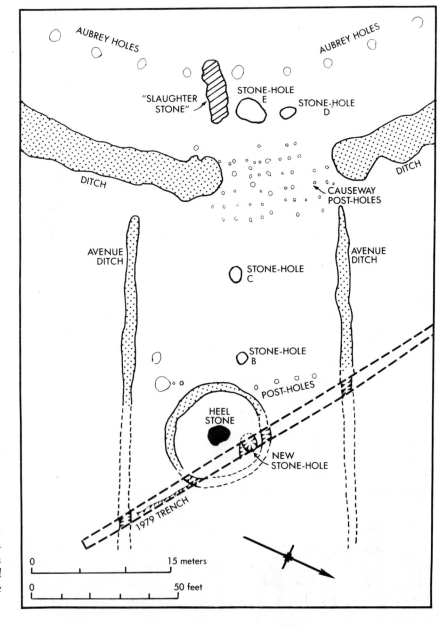

AUBREY HOLES

AUBREY HOLES

"SLAUGHTER STONE"

STONE-HOLE E

STONE-HOLE D

DITCH

DITCH

CAUSEWAY POST-HOLES

AVENUE DITCH

AVENUE DITCH

STONE-HOLE C

STONE-HOLE B

POST-HOLES

HEEL STONE

NEW STONE-HOLE

1979 TRENCH

0   15 meters

0   50 feet

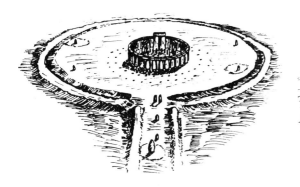

leading British prehistorian, R. J. C. Atkinson, and his colleagues during the 1950s, they established that there were actually several Stonehenges. Like a cathedral, it was built and rebuilt on the same spot over a span of nearly two thousand years (starting roughly around 3000 B.C.). Successive construction work in earth, wood, and stone destroyed much evidence that would allow us to trace astronomical alignments with confidence. The gaps in our knowledge continually frustrate attempts to evaluate the different astronomical theories. One small scrap of fresh archaeological information could substantially change ideas about the astronomy of Stonehenge.

This was demonstrated dramatically in the summer of 1979, when officials of the British Post Office were preparing to lay a telephone cable beside the modern tarred road that passes close to the Heel Stone. A team of archaeologists, led by Michael Pitts of the Avebury Museum, was hastily summoned to examine the narrow strip of chalk subsoil that was to be destroyed by the burying of the cable. Although this was only a limited excavation, a remarkable discovery was made: another stone had once stood upright about ten feet north of the present location of the Heel Stone. The base of the stone had left an impression in the chalk, about half of which was uncovered by the diggers, and it appears to have been substantial in size. The stone had been removed, however, and the pit in which it stood had been filled in again. It is impossible to tell exactly when this happened, but it seems likely to have been early in the history of the monument.

Although we cannot be certain that the new stone and the Heel Stone were originally set out as twins—a flanking pair—the possibility is intriguing. The new stone would have appeared to the left of the Heel Stone when seen from the center of the circle; perhaps, at the outset of the long development of Stonehenge, dawn was meant to break between a "gateway" of stone rather than over the tip of the Heel Stone itself.

This new discovery demonstrates the incompleteness of our knowledge and the difficulty of assessing the astronomical claims. Could it all be wishful thinking, this urge to create an astronomical observatory from the ruined stones? Was the Heel Stone actually erected for some unknown religious purpose, and does it just happen to line up roughly with the sunrise?

When half finished, the bluestone circle was dismantled to make way for the famous ring of sarsen stones transported from the Avebury region, twenty miles north, top left. In a later phase (c. 1550 B.C., right) the bluestones were raised in a horseshoe design within the sarsen circle, much as we see them today. The avenue was lengthened in about 1100 B.C., suggesting the continuing importance of Stonehenge. Many features shown on these drawings are conjectural.

THE ERRORS OF STONEHENGE

If it could be shown that many other astronomical alignments are present at Stonehenge, the likelihood of coincidence would be small. In fact, in his first article published in the scientific journal *Nature* in 1963, Hawkins presented a list of twelve solar and twelve lunar lines which he had found from careful measurements of a plan of the monument. These twenty-four significant lines emerged as a result of his examination of about fifty pairs of directions drawn between stones and pits. It seemed improbable to Hawkins that such a high score of astronomical "hits" could be due to chance.

But other scholars have disagreed, notably R. J. C. Atkinson and the Scottish astronomer Douglas Heggie. They have criticized Hawkins' work, pointing out that some of his alignments could be an accident resulting from the distinctive layout of Stonehenge. They believe that the grand total of alignments should actually be much greater than the fifty pairs he selected; this would make the statistics of Stonehenge far less convincing. Their criticisms illustrate once more the problem of assessing such a complicated monument. Although statistics offer an impartial means of testing the astronomical theories, opinions differ on exactly how the tests should be set up in the first place.

Stonehenge from the air.

The leading authority on the archaeology of Stonehenge is Professor Richard Atkinson, seen here in 1981 explaining the position of the so-called "Slaughter Stone" close to the axis of the monument.

There is a further problem in evaluating Hawkins' remarkable work: the builders of Stonehenge were apparently as careless in marking most of their observations as they were in aligning the Heel Stone with the midsummer sunrise. A degree or two of accuracy seemed to be sufficient. Atkinson experimented with a pair of sticks and thought he could lay out alignments with twenty-four times more accuracy than the Stonehenge observers had supposedly done.

Responding to such criticisms, Hawkins supervised a new and more exact survey of the monument, using air photographs taken for that purpose in 1965. As a result, the errors in the astronomical alignments were reduced to an average value of about plus or minus one degree. This is still a much greater error than we would expect the builders to have made, using good eyesight and a steady hand as they hauled the stone markers upright. (A third of one degree should have been possible.) But were their standards as exacting as our own?

Despite Hawkins' belief in the persistence and sophistication of the builders' interests in the sky, he considered it unlikely that they should ever have striven for more than a degree or so of accuracy in erecting substantial stones. If they "missed" the sun and moon by a diameter or two here and there, such errors were unimportant. They reflected only the practical difficulties of positioning the large stones, which presumably had replaced earlier, more precise markers of wood. In Hawkins' view, the prehistoric astronomers were not driven by a passion for perfect results like a team of scientists seeking out new discoveries. Instead, Stonehenge was a symbolic monument recording an ancient awareness of the rhythms of the sun and moon, inspired by religious beliefs that would probably remain obscure and elusive. While Hawkins became famous for using a computer to "decode" the alignments, he never claimed that Stonehenge itself was a high-precision "machine."

Even if we suppose that the Heel Stone had been positioned *exactly* in line with the ancient sunrise, it could never have served as an accurate calendar marker. To understand this, it is necessary

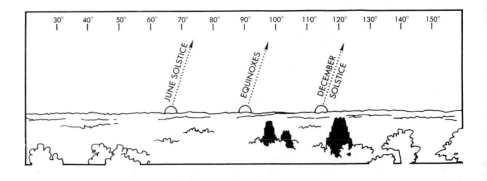

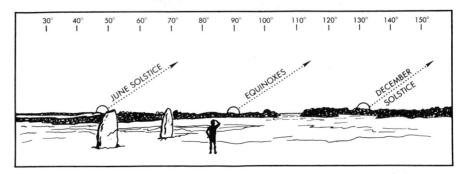

The movement of the sun along the horizon through the seasons as viewed from the tropics, top, and a northern latitude, bottom. In both cases, this movement is not uniform: the daily change in the sunrise position becomes progressively smaller with the approach of midsummer and midwinter. But in the northern latitudes the changes are more spread out along the horizon, and are more obvious to the observer.

to visualize the change in the sun's rising point along the horizon each morning. During the first half of the year, the daily position of sunrise advances progressively farther to the north; after midsummer it retreats southward again until the shortest day of the year is reached at midwinter.

The swing of the sun during the course of the year is not an exactly steady rhythm, however. A little before both midsummer and midwinter (also known as the "solstices"), the sun "slows down" relative to the horizon so that the distance between each successive dawn decreases. At the latitude of Stonehenge, the sun will appear to rise and set more or less at the same spot on the horizon for about a week around the time of the solstices.

This means that a crude pointer like the Heel Stone, set only a short distance away from the observer's eye, is useless for establishing a calendar that is required to be accurate to within a single day. In early June, we would note the sunrise creeping steadily northward toward the marker, but when it reached the stone there would be no day-to-day movement for about a week or so. Even if visibility conditions were perfect, we would still not be able to distinguish the difference between one day's sunrise and the next until the slow southward shift began again. With such alignments, our knowledge of the length of the year and of the exact date of midsummer or midwinter would only be approximate, within a few days on either side of the true solstices.

Did the Stonehenge builders actually strive for better accuracy than this? It is possible that they *deliberately* positioned the Heel Stone so that it stood a little short of the extreme point of sunrise. They would have found it easier to judge when the sun rose on two dates a few days before and after the solstice. If they had then halved the time between the two observations they would have found out the day when the sun had actually reached the end of its swing. In other words—if this interpretation is correct—the errors in the setting up of Stonehenge were not errors at all. They were not due to clumsiness or carelessness on the part of the builders, but were necessary for a clever method of observation.

Unfortunately, attractive as this idea sounds, not all of the actual errors involved in the solar and lunar alignments fall into the orderly pattern we would expect if the observers were intentionally aiming a little off-target. The inaccuracies in the positioning of stones and posts cannot be explained away by any one simple argument.

The "decoding" of Stonehenge by Gerald S. Hawkins was an important pioneering step in the study of ancient astronomy, and it has led to a general acceptance of the idea that astronomy at some level was indeed practiced at Stonehenge. But as we have seen, it is difficult to establish the extent of this activity by objective statistical tests. We cannot be sure how far the observations were pursued or what, precisely, motivated the efforts of the Stonehenge skywatchers.

But fancy speculations that call for the existence of prehistoric Einsteins should make us pause for thought. The errors that seem to be involved in many alignments at Stonehenge suggest a different picture. Our modern emphasis on precision may not have existed in that culture; instead, the Stonehenge people may have feared the power of the sky gods and sought to appease them by tracing their mysterious cycles of movement, rather than pursuing a scientific program of research as we would understand it today.

Were the narrow gaps between the Stonehenge archways intended for astronomical viewing, as Gerald S. Hawkins has suggested? Unfortunately, the ruined state of the monument today hinders any clear assessment of this theory.

4. Death and the Sun in Ancient Britain

NEWGRANGE: THE SUN REBORN

In the minds of the people of prehistoric Britain, burial customs were linked to beliefs about the sun and moon. At midwinter, traditionally the season of death and renewal, these beliefs seem to have been felt and expressed most keenly.

A little after sunrise on the shortest day of the year, a thin beam of light penetrates the stone passageway of the most impressive ancient tomb in Europe. This is the monument known as Newgrange, located on a fertile ridge overlooking the River Boyne in eastern Ireland, about twenty-eight miles north of Dublin. The ray of sunlight shines deeply down the passageway, even though it follows a slightly twisting course toward the heart of the tomb. Eventually, the sun creeps all the way to the far end of the central burial chamber, a distance of about eighty feet from the entrance. Finally the beam touches the front edge of a stone basin probably used to hold the cremated remains of several individuals at least five thousand years ago. The basin is overlooked by an intricate design of triple spirals carved on the tomb wall above. For a quarter of an hour or so, the innermost sanctuary is softly illuminated, and then, as the sunlight retreats across the chamber floor, darkness reigns again in the empty tomb. This theatrical event is repeated for about a week before and after the winter solstice, and then for another year the chamber returns to its lifeless gloom.

The construction of Newgrange, which took place around 3300 B.C., was an extraordinary engineering operation. At least four hundred giant glacial boulders were carefully selected and assembled to make the passageway, chamber, and encircling kerb-ring.

The megalithic tomb of Newgrange, Ireland, erected in about 3300 B.C. Note the surrounding ring of stones and the almost vertical front façade of white quartz. This was restored following excavations by Michael J. O'Kelly, who discovered evidence that the façade must have collapsed shortly after its original erection.

The roof was a daringly conceived vault of overhanging or corbeled slabs, each one piled on the next until the final covering stone was nearly twenty feet above the chamber floor. The tomb was buried under an immense mound composed of neat layers of turf and water-rolled pebbles. The competence of the builders was only fully appreciated after years of excavation and restoration directed by the Professor of Archaeology at Cork University, Michael J. O'Kelly, and his wife, Claire. Astonished at the discovery of an ingenious system of carved drainage channels to divert rainwater off the top surface of the roof slabs and hence away from the interior of the tomb, Claire O'Kelly commented that "it becomes more and more evident that the building of Newgrange was no trial-and-error, hit-and-miss affair, but the work of practised builders who fully appreciated the factors which would best assure long life for their monument."[11]

Such practical skills and concerns were combined with a remarkable symbolic plan, demanding that an astronomical alignment run right through the entire length of the tomb structure. This involved considerable ingenuity. Originally, the tomb was arranged so that light did *not* shine directly through the main entrance, which was sealed by a blocking stone and possibly also by rubble from the mound. But a boxlike arrangement of two slabs angled above the entrance forms a kind of slot, much too narrow for anyone to crawl through, but positioned to frame the disc of the rising winter sun.

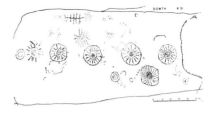

While a number of Irish megalithic tombs are richly decorated with engravings, few of the carved symbols convey obvious astronomical meaning. The suns carved on an exterior stone at the tomb of Dowth, close to Newgrange, are one exception.

The entrance to the Newgrange passage, with the roof-box opening above it and the magnificently carved entrance stone in the foreground.

The spectacle of the sun shining through this narrow gap has convinced many eyewitnesses that the Newgrange alignment cannot be accidental. As Claire O'Kelly writes,

> . . . it is difficult to remain sceptical once one has actually seen, as I have, the thin thread of sunlight striking in along the passage at this most dismal of all times of the year until the dark of the chamber begins to disperse and more and more of it becomes visible as the sun rises and the light strengthens. Upon looking outwards towards the entrance, one sees the ball of the sun framed dramatically in the slit of the roof-box and one realizes that in the whole course of the year this brief spell is the only period when daylight has sway over the darkness of the tomb.[12]

As if to confirm the deliberate orientation of the tomb, the entrances of four much smaller "satellite" graves erected in the shadow of the huge mound also face southeast toward the midwinter sunrise. Inside the ruined chamber of one, excavators found a stone basin that must once have held cremated bones, like the burial deposits discovered in other less disturbed tombs of this type. On its surface the stone basin still displays the carving of a circle and rays resembling the sun.

Of all the prehistoric sites in Britain, Newgrange reminds us most insistently of the connection between beliefs about the afterlife and about the sun. It was not an observatory, for the alignment was too inaccurate to provide useful information for a calendar. Neither does it seem likely that people regularly entered the chamber to watch the sunrise each winter as part of a religious ceremony. If this had been the intention of the builders, they would surely have arranged for sunlight to penetrate through the opened main entranceway, rather than troubling to design the special slot above it. Instead, it looks as if the astronomical event was sacred, hidden from the eyes of men and women. The sun's rays were intended for the ashes of the dead, not the calculations of the living.

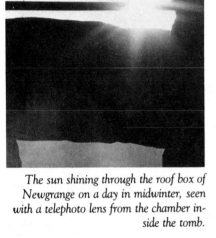

The sun shining through the roof box of Newgrange on a day in midwinter, seen with a telephoto lens from the chamber inside the tomb.

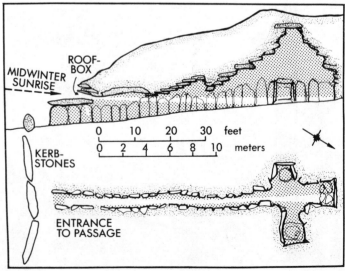

A simplified plan of Newgrange, showing the path of the sunrise ray.

The splendor of Newgrange seems to have ensured that the influence of the builders' beliefs was widespread and long-lasting. For example, the plan of another impressive tomb known as Maes Howe is similar to that of Newgrange, even though it was built far away in the Orkney Islands and dates to about five hundred years later, around 2500 B.C. As in the case of the Irish monument, the main entrance of Maes Howe was originally sealed by a blocking stone. This carefully shaped stone was not quite high enough to cover up the entire mouth of the passage, however; it would have left a narrow gap like a letterbox, about 3 feet wide and 1½ feet high. Like the roof-box at Newgrange, this slot seems to have been devised to allow midwinter sunlight to enter the tomb, although at Maes Howe the alignment has not yet been properly investigated to establish exactly how far the rays of the setting sun penetrate around the time of the solstice. In any case, just as at Newgrange, Maes Howe presents evidence of an astronomical alignment that probably had a symbolic importance, but the tomb itself is unlikely to have been an observatory for regular skywatching.

More than a thousand years after the construction of Newgrange, prehistoric people in Scotland were raising burial monuments that still seemed to reproduce some important features of the Irish tomb. These monuments are the distinctive "kerb-cairns," found in the regions of Perthshire, Argyll, and Aberdeenshire. The name refers to the kerb or boundary of each monument, which consists of a ring of massive stones lying on their sides with ends touching. There are no carvings on these stones as there are at Newgrange, yet there are other similarities. At the site of Kintraw on the west coast of Scotland, for example,

A DOORWAY TO THE DEAD

A view from inside the chamber of the megalithic tomb of Maes Howe, in the Scottish Orkney Islands, looking in the direction of the entrance passageway and toward the sunset in midwinter.

A general view of the ruins of Kintraw in Argyll, Scotland. The remains of the small cairn lie in the foreground, with the standing stone and large cairn beyond.

The "false entrance" built against the southwest side of the large Kintraw cairn. A photo taken by the excavator D. D. A. Simpson during his dig of 1959-1960.

a pair of typical kerb-cairns were excavated in 1959 and 1960. Around the edges of the big cairn, the archaeologist Derek Simpson found many fragments of quartz and a few of crystal, suggesting that originally the mound had been covered with a sparkling white coating of rock, again similar to Newgrange. Underneath the mound material, both cairns at Kintraw concealed little boxlike stone graves. One of them contained a few scraps of cremated bone and burnt wood. Here were the burials that must have played an important part in the beliefs of the cairn builders.

The most curious find at Kintraw was an arrangement of stones on the southwest edge of the large cairn. At this spot there stood a "doorway" of two tall upright slabs, set securely in holes dug into the earth. But this was a doorway that led nowhere, for the "entrance" between the slabs was blocked by a great kerbstone and was also filled up with smaller rocks interspersed with charcoal. As if to draw attention to this peculiar "entrance," a huge stone lay flat on the ground just in front of it, with thick scatters of quartz all around. This false entrance must have resembled the blocked passages of earlier tombs such as Maes Howe and Newgrange.

Why was this symbolic "doorway" positioned on the southwest rim of the big cairn? When "viewed" from the center of the cairn, this direction corresponds to a lunar alignment. It is extraordinary to imagine an astronomical sight line that nobody could have observed after the mound was built, yet that is what the arrangement of stones at Kintraw appears to be. Perhaps the setting of the moon was "watched" by the spirits of the dead inside the cairn—a view not intended for the living.

As for the little cairn, it, too, seems to have been lined up in a way that could have served no practical type of observation. A great twelve-foot pillar of stone stands close by, very roughly in the direction of midsummer sunrise. The glittering white mass of the large cairn would, in fact, have blocked the view of the sun (though perhaps the large cairn was a later addition to the site).

The important point is that the builders of *both* cairns at Kintraw appear to have been preoccupied with astronomy as part of the correct ritual procedure for setting up monuments to the dead. Once this had been accomplished, the astronomy was useless, unless the builders thought that the sun and moon lines continued to benefit the dead whose ashes lay secure under the rubble. Summing up the evidence of Scottish sites such as Kintraw, all raised a little after 2000 B.C., the archaeologist Aubrey Burl writes that "the kerb-cairns are unlikely to have been observatories in the sense that they were used for recurring observations. They were religious monuments incorporating astronomical alignments in their design."[13]

A very different view of the Kintraw ruins was proposed by Thom in the late 1960s. In contrast to the archaeological evidence for crude, symbolic alignments, Thom believed that the cairns were the remains of a highly accurate solar observatory. His theory involves the dramatic landscape surrounding the site, which is located on an elevated plateau overlooking the shores of Loch Craignish. If a visitor to Kintraw succeeds in dodging the persistent downpours of rain, it is possible to catch a glimpse of the high peaks of the island of Jura on the far horizon, nearly thirty miles away to the southwest. This is also the place where the sun sets at midwinter.

The controversial sight line from the platform at Kintraw toward the distant Paps of Jura and the midwinter sunset.

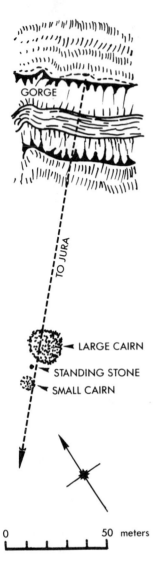

GORGE

TO JURA

← LARGE CAIRN

← STANDING STONE

← SMALL CAIRN

0   50 meters

A sketch plan of the alignment at Kintraw.

Now suppose that an ancient observer positioned near the Kintraw cairns had tried to line up the edge of the mountains on Jura, like a gun sight, with the last flash of the winter sun. This alignment was so long and so sensitive that the observer would find that he or she had to stand in a different spot each evening. In theory, it would have been possible to detect the tiny change in the sun's movement only a day before or after the solstice (equal to about one three-hundredth of a degree). So the alignment could have been used to correct a precise calendar, accurate to a single day. The tally of days in the year could be checked each winter if the weather permitted. It would be reasonable for us to compare such an achievement of precise, orderly time measurement with the aspirations and efforts of scientists today.

However, there is a serious objection to the theory: the kerb-cairns of Kintraw are situated *where the distant mountains of Jura are invisible.* An intervening ridge known as Dun Arnal juts out into the loch, shutting off the view of the far peaks and of the setting midwinter sun. The alignment nevertheless seems to be a near miss; if the viewpoint of the observer was a little higher up, then the horizon would come into view. Thom believes that the summit of the big cairn would have provided the necessary extra elevation for the skywatcher, and so, according to his theory, the monument was built to serve as a phenomenally accurate solstice observation point.

If we suppose that this theory is correct, how did the builders of the observatory decide where to locate their cairn in the first place? Since the view of the mountains of Jura was blocked by Dun Arnal, there was no knowing where the cairn should be positioned. The obvious answer was to climb the nearest hill behind the site, do the first observations there, and so work out the right spot for the cairn down at the lower level.

Thom searched for evidence to confirm this idea and discovered a platform at a suitable place about fourteen feet higher up than the ground beside the cairns, apparently cut into the steep hillside. From this platform the outline of Jura could just—and only just—be made out above the thick foliage covering the ridge of Dun Arnal. Standing here at midwinter in about 1800 B.C., an observer would have seen a momentary flash of the dying sun as it passed below the cleft in the mountains.

Was the terrace on the hillside a natural or manmade feature? Here was a crucial test for the entire theory. If traces of ancient activity were discovered there, it would be hard to deny that prehistoric people had actually carried out the ingenious observing technique proposed by Thom.

High expectations were raised when Euan MacKie, an archaeologist at the Hunterian Museum in Glasgow, began to excavate the platform in the summer of 1970. At the exact point predicted by Thom's theory, MacKie uncovered a pair of boulders and a level layer of stones running for at least twenty feet toward the southeast. Unfortunately, no artificial remains of any kind, not even a single shattered fragment of pottery, were found among the "paving" stones. Furthermore, the opinion of geologists is still divided on whether the stones represent a manmade surface or a natural feature. There is a distinct possibility that the platform is actually a modern cow track leading from Kintraw Farm to pastureland nearby.

During several successive visits to Kintraw, I became convinced that the narrow platform was an impractical place for the kind of observations demanded by Thom's theory. The view of the distant horizon is still unsatisfactory. On some occasions the cleft in the peaks of Jura was visible just above Dun Arnal, but on other visits the ridge blocked the view, probably because of changes in atmospheric conditions along the thirty-mile sight line. Such variations in refraction would certainly have presented serious difficulties to an ancient astronomer who attempted to carry out the type of observing envisaged by Thom.

Moreover, on evenings immediately before and after the solstice, the observer would have been required to stand wherever the last flash of the sunset could be glimpsed; these positions would all be considerably to the left of the boulders (about 19 feet the first day before or after; then roughly 75 feet on the second day; next day, 170, and so on). It is obvious from the photograph that even a slight shift toward the left will cause the peaks to disappear once again behind the crest of Dun Arnal.

These factors indicate that Kintraw was an impractical place to carry out precise solstice observations. The layout of the ruined cairns at the site may reflect an awareness of the solar and lunar movements, but this awareness seems to have been connected with beliefs about the dead, rather than with efforts to establish an accurate calendar.

THE CAIRN OF THE ANCESTORS

The arguments for accurate astronomy at Kintraw would be more convincing if they could be supported by evidence from other sites in western Scotland. In fact, Thom has proposed that another high-precision solar observatory existed at the site of Ballochroy, a grassy field overlooking the sea some thirty-five miles south of Kintraw.

This, too, was a burial monument: the prehistoric builders deposited the remains of the dead inside a simple square box constructed of stone slabs, then piled rubble over the grave. This mound must have been substantial enough to give rise to its local Gaelic name, Carnmore ("the big cairn"). Sometime after a sketch of the site was recorded around 1700 A.D., the cairn rubble was stripped away as a convenient source of building material for the nearby stone walls. All that is left of the site today is the empty burial box and a row of three upright monoliths. However, these remains are particularly significant for anyone interested in ancient astronomy.

The three stones stand in a line which, if projected to the

The three standing stones at Ballochroy, Argyll, appear to be turned toward the Paps of Jura. But did the erectors observe midsummer sunset over the Paps with high accuracy?

southwest, runs through the stone box of Carnmore about 120 feet away. This is also the direction of midwinter sunset, so we can imagine how the last feeble glow of the sun could have been watched from the stones, sinking behind the big cairn. Perhaps it was thought that the power of the sun actually did "enter" the cairn, restoring life to buried ancestors. Surely similar thoughts had led the builders of the much more elaborate tomb of Newgrange to ensure that rays of the winter sun really could penetrate the sanctuary of the dead. In any case, the probability that the astronomical alignment is a deliberate one seems high, since it is indicated by four well-separated features along its length.

The latitude of the British Isles is such that sight lines to the summer and winter solstices meet at angles close to ninety degrees. It is tempting to speculate that the erectors of the Ballochroy stones were aware of this fact, for they turned the flat faces of two of the three stones at right angles to the main alignment that ran through the big cairn. (The third stone, the tallest at the site, is roughly square in section and so cannot be said to point anywhere in particular.) It seems, then, as if the builders were watching sunset at midsummer as well as at midwinter.

This conclusion is a cautious one, since the flat faces of the stones provide a much less definite indication than the long line stretching toward the southwest. There was no burial monument for the summer sun to sink behind, but Jura's striking succession of mountain peaks did rise to the northeast. Perhaps we can

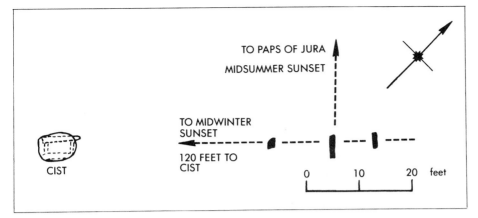

TO PAPS OF JURA

MIDSUMMER SUNSET

TO MIDWINTER
SUNSET

120 FEET TO
CIST

CIST

0   10   20   feet

Plan of Ballochroy, traced from Thom's survey published in 1954.

imagine the peaks of Jura as "sacred mountains"; was the sun's "return" to them at midsummer a time of religious significance?

Thom interprets the stones of Ballochroy differently. He argues that prehistoric astronomers could have lined up the last flash of the sun with a distant mountain slope, exactly as they could have done from an elevated spot at Kintraw. The strength of his argument lies in the fact that the alignments at *both* sites run to the *same* range of peaks on Jura. Furthermore, the calculated figures agree exactly if both sites are assumed to date to about 1800 B.C. Armed with this evidence, supporters of Thom's theory suggest that Ballochroy and Kintraw were set out together as complementary observatories. One gave the exact day of midsummer, the other midwinter. Both contributed to the regulation of an astonishingly precise calendar by ancient observers who might indeed deserve the title of "scientists."

The appeal of this argument has convinced many, yet there are serious doubts about the long-distance sight line at Kintraw. There are just as many problems at Ballochroy. The tallest stone points nowhere. The shortest stone *is* turned toward Jura, but if one picks out a mountain slope in this direction and measures it, no astronomical alignment will fit. The theory depends on selecting only the middle stone and claiming that its flat face was intentionally aimed toward the correct slope on Jura. This is too much to claim for such a crude indicator as the side of a standing stone.

The most reasonable explanation is that *both* smaller stones *were* directed at the setting midsummer sun, but only in a general, imprecise fashion. This was perhaps a relic of the piety rather than the shrewdness of the observers.

As for the long sight line to the southwest—the convincing layout of the three stones and the Carnmore grave toward the winter sun—the evidence surely weighs against the observing technique proposed by Thom. The cairn must have blocked the view of the distant horizon. While the alignment was of symbolic importance to the builders, it could not have been used for repeated, accurate sunwatching.

Of course, it can be argued that the cairn was built exactly in line with the stones at a later date, after the observatory had fallen into disuse. But unless evidence for this becomes available, it is reasonable to prefer the simpler theory, and to reject the idea that Ballochroy was operated as an accurate solar observatory.

These two Scottish sites illustrate the dilemma faced by anyone interested in the astronomy of the megaliths. If we believe in a precise Stone Age science, it is necessary to find evidence that sensitive, long-distance sight lines really did exist. This is difficult to do if we rely only on the surface appearance of sites as we see them today, disfigured by centuries of erosion and neglect. Even when excavation was tried at Kintraw, the evidence was elusive and unsatisfactory.

On the other hand, archaeologists have collected information based on studies not just at one but at groups of similar sites. This provides evidence of inaccurate, short-distance astronomical alignments of a "ritual" rather than "scientific" character. It leads us to consider fascinating but unprovable ideas such as a "door-way" between the rays of the moon and the ashes of the dead. If we take such speculations seriously, then we transform the megalith builders from narrow-minded technologists into much more interesting characters, with heads full of fears and symbols as well as skills and knowledge.

5. The Moon and the Megaliths

The evidence of two sites cannot prove or disprove a whole theory. It is not very useful to pick holes in Thom's work simply because two examples seem doubtful out of the hundreds he so patiently surveyed and studied. A better approach is to consider a large number of alignments drawn from many different standing stone sites, as Thom himself has done. With a generous sample of sight lines, we can then use statistical calculations to judge whether the alignments were deliberately intended by the original builders or whether they are merely an accident. The possibilty of coincidence should not be underestimated, since the plans of some megalithic ruins offer dozens of potential lines, while the positions of the sun, moon, and stars present a wide range of astronomical "targets."

The task of statistical analysis is obviously a complicated one, reserved for those with mathematical training. Yet the outcome of the calculations is vital to anyone interested in whether or not astronomy was practiced in Britain four thousand years ago.

It was Thom who made the first serious attempt, in 1954, to compile a collection of accurately measured alignments. Since this study appeared in a specialized journal read by astronomers, at the time it attracted little notice from archaeologists. Nevertheless, it was a remarkable piece of work. Among other theories, Thom produced strong evidence that about a dozen megalithic sites were oriented to the rising or setting points of the sun at the solstices. This evidence was confirmed by additional sun lines presented in a paper he wrote the following year. These early works of Thom, which continue to impress mathematicians and astronomers, contain only cautious claims about the accuracy of the observations conducted by the megalith builders. The paper published in 1955 merely hints at the possible existence of alignments to the moon.

By the time Thom's first book, *Megalithic Sites in Britain*, appeared in 1967, he had assembled a list of over 250 lines drawn from surveys of nearly 150 sites. They represented a great variety of monuments, including stone circles and single standing stones, tombs and cairns, pairs and rows of stones, and other types of markers. The number of possible celestial targets had also multiplied; now Thom included alignments to the moon, the stars, and solar calendar dates in his discussion. Several scholars have pointed out that his standards for selecting lines gradually became less strict as his belief in the existence of a precise Stone Age science grew. From 1967 onwards, it was increasingly difficult to evaluate the part played by chance in his work.

Nevertheless, as Thom became a well-known figure during the 1970s, his theories continued to be a source of original, stimulating ideas. Even if the statistical basis was not as reliable as it was

THE LUNAR RHYTHM

In his recent work on megalithic lunar alignments, Alexander Thom has collaborated with his son Archie, here seen at Avebury in 1981.

for the earlier solstice alignments, Thom's later work raised the important possibility that lunar alignments might exist at other sites beside Stonehenge.

The moon is an important test of the abilities of ancient astronomers because its movements are more complex and demand more skillful observation than those of the sun. In urban surroundings we may notice the different phases of the moon from time to time, but the place in which it appears in the night sky can seem quite haphazard from one week to the next.

Actually, the movements of the moon are perfectly regular, as most people not hemmed in by tall buildings or dazzled by city lights are well aware. The nightly change in its position is difficult for us to follow only because it is more rapid than the daily progress of the sun. Instead of taking a whole six months to sweep across the heavens as the sun does between midsummer and midwinter, the moon swings between the limits of its path in only *two weeks*. Just as the sun reaches its farthest rising points on the northern horizon at midsummer and on the southern horizon at midwinter, so, too, there are northern and southern boundaries to the moon's movement. If we were to set up stone or wooden markers and watch the moon every night, we would find that each month it never rose beyond a certain point on the skyline.

Besides its monthly movement to and fro, however, a further subtle variation affects the moon. The northern and southern boundaries change very slowly over a cycle of about 18½ years in length (to be exact, 18.61 years). To visualize the cycle, we can imagine the moon's track sweeping out a broad area of sky every nine years, to return about nine years later to a narrower band of movement.

Let us suppose that the extreme rising point of the moon every month was watched and lined up with a post or stone, or a distant mountaintop, over a period lasting for more than a generation. With persistence and luck, the observers would notice the monthly marker very gradually shifting its location. Discovering this change was one step; realizing that it was part of a regular cycle, back and forth, stretching over more than half the average person's lifetime, was another.

Diagram of the western horizon, showing the extreme setting points of the moon over its 18½-year cycle.

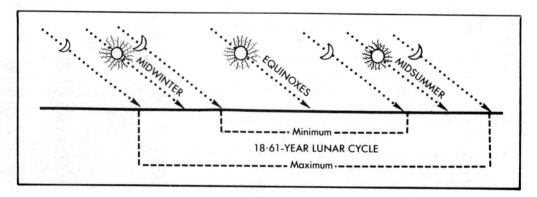

We should not underestimate the difficulty of acquiring this added knowledge of the lunar movements. It would have been hard to detect the change just from one month to the next because it was so tiny—a mere fraction of the width of the moon's disc itself. Furthermore, the phases of the moon do not always coincide with its movements through the sky each month. The full moon would be easy to spot and follow as it rose behind a mountain or a stone, but a thin crescent would not. The moon might also rise at one of its limits during the day and so be completely invisible near the horizon.

Only around midwinter did the full moon appear at the same time as it rose at its farthest point in the northern sky. If this full moon next to the shortest day of the year had been watched systematically, then slowly the great $18\frac{1}{2}$-year cycle would have revealed itself. The slight variation in the moon's behavior would probably have fascinated and perplexed the observers, especially if the sun and moon were regarded as living forces to be feared and worshipped.

Incidentally, the inhabitants of the far north of Scotland were more likely to know about this long-term cycle of the moon than people in the south. For example, at the latitude of 62° north in the Shetland Islands, the lunar equivalent of the Arctic "midnight sun" can occur. About every $18\frac{1}{2}$ years, the dramatic nights arrive when the moon never sets but merely brushes the northern skyline. An observant prehistoric farmer would surely have known of the moon's slow swing.

Farther south, the motion was not so obvious, but even today people in remote rural areas are often aware of subtle changes in the sky. Recently, while investigating stone circles in southwest Ireland, the archaeologist John Barber learned from local farmers that there was a period called the *Duibhré* when (once every eighteen or nineteen years) the moon is so low in the sky that it fails to rise above the surrounding mountains. A consciousness of the long-term lunar motion is therefore not beyond the faculties of a watchful farmer.

But *did* prehistoric people know of the moon's cycle? And did they watch it with scientific accuracy?

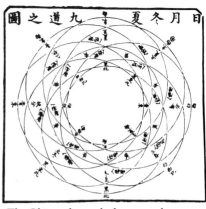

The Chinese knew the long-term lunar cycle as the "Nine Roads of the Moon", and represented it in this symbolic design from the late Ch'ing period.

MOON RITES IN ANCIENT SCOTLAND

The inhabitants of northeast Scotland seem to have recognized the long swing of the moon at least five thousand years ago, to judge from a remarkable study published by Aubrey Burl in 1981. A group of circular tombs and cairns known as the Clava group, clustered near the modern town of Inverness, have attracted attention for many years because of their unusual and impressive construction. As in the case of the Irish tomb of Newgrange, most of the Clava cairns are surrounded by a ring of standing stones. The height of the stones rises regularly toward the southeast; it is on this side of the monument that the passageway of the tomb is always located. So few human remains have been discovered inside the chambers of the Clava sites that Burl suggests they may have been built principally as shrines rather than as burial places.

When he plotted the directions in which the tomb passages faced, Burl discovered that nearly all of them fell neatly within the arc swept out by the setting points of the moon in the course of its 18½-year cycle. The sites, however, could not have been lunar observatories; the view from inside the chamber along the passage is very restricted, and in the case of half a dozen of the best-preserved examples, it is impossible to see the horizon at all. Once again, we encounter astronomical alignments which were part of the ritualistic architecture of ancient monuments, yet could not have been used for repeated observations by the people who planned them.

By about 2500 B.C., the Clava cairns had probably already fallen into disuse. Not far away, in the present-day "whisky country" of central Aberdeenshire, prehistoric farmers then began to erect another distinctive type of monument.

In general, most stone circles frustrate the astronomer because so many possible directions are indicated by the positions of the stones. It is impossible to single out a line drawn between one pair of stones in preference to another. But about eighty "recumbent" stone circles in northeast Scotland form a highly exceptional group. They are called "recumbent" because a huge slab of stone was deliberately laid on its side at the southern edge of each circle. Further attention was drawn to this block by the arrangement of the other stones, all upright, which rise regularly in height as they approach the recumbent, so that it is flanked by two tall pillars on either side. This gradation in height suggests that the builders followed some of the traditions of the earlier Clava cairn people.

A chambered tomb at the Neolithic cemetery of Balnuaran of Clava, Inverness, dating from about 3000 B.C. The tomb contains a passage leading to a circular chamber about 13 feet across, where diggers in 1854 uncovered "a few bones." The cairn is surrounded by eleven standing stones.

Burl, who spent much of the 1970s studying these neglected monuments, believes that most circles were small enough to have been built by a single farming family. In some cases, however, the massive recumbent must have been moved with the help of neighbors and, presumably, such aids as wooden sledges and rollers. At Old Keig, Aberdeenshire, the mighty fifty-ton stone was heaved from six miles away, a feat involving the muscles of perhaps two hundred or more people.

Like other researchers before him, Burl was at first mystified by the meaning of the recumbent and why it was positioned so consistently between the southwest and south-southeast points of the compass. Although the location picked for the circle always commanded a good view of the horizon, the orientation of most recumbents did not fall within the risings and settings of either the sun or the moon.

But what of the full moon, once it was riding high in the midsummer sky? This would surely have formed a dramatic backdrop to whatever nocturnal dances, ceremonies, or gatherings may have been staged within the ring. About every 18½ years, the moon would sweep past, appreciably nearer to the horizon, in a few cases actually appearing to "enter" the "gateway" formed by the recumbent and its tall flanking pillars. What beliefs could lie behind this union of the circle with the moon?

It is clear, at least, that the recumbent stone circles were not observatories. To gain *precise* knowledge of how the moon was moving, it would have been necessary to watch its rising and setting positions, with which the builders were apparently unconcerned. Instead, they located nearly every recumbent "underneath" the moon's path as it swept out its maximum arc across the sky. The purpose of this positioning could only have been for religious symbolism, not science.

Presumably the builders' observances reached a peak every 18½

Air view of the recumbent stone circle of Loanhead of Daviot, Aberdeenshire, erected around 2500 B.C. A funeral pyre blazed at the site before the central ring-shaped cairn of small stones was added. Finally, about five pounds of human bone (including many infant skull fragments) were deposited in a pit at the center.

Above, like many other recumbent stone circles in Aberdeenshire, Old Keig was arranged so that the immense recumbent stone lay at the south–southeast with its top surface almost perfectly level. At Strichen, right, a recumbent stone circle was incorrectly restored during the early nineteenth century by Lord Lovat of Strichen House, whose mansion appears in the background.

From 1979 to 1983 the site was completely excavated by Aubrey Burl, Ian Hampsher-Monk and Philip Abramson. In their efforts to restore the site correctly, volunteers at Strichen, below, discovered the most efficient means of moving stones was to use logs as a kind of sledge, dragged over a slippery trail of dry straw.

years as the full moon at midsummer sank lower in the sky. A powerful kind of magic may have been practiced on these occasions, since the interior of the monument usually contained pits full of charcoal, shattered pottery, and white flecks of burnt human bone (sometimes of young children). The circles were not, however, burial grounds in the ordinary sense. The cremations were only "tokens," usually representing a few individuals, with just a small portion of the total bodily remains present. Human bones must have lent sanctity to the sites.

In addition, fragments of quartz were customarily scattered around the recumbent stone. The milky white quartz may perhaps have represented "moonstone," intended to draw the lunar influence down into the circle through the pillared "gateway" formed by the recumbent and its flanking uprights. (As we saw in the previous chapter, similar features excavated at the large cairn at Kintraw were also apparently linked with the moon. The kerb-cairns and recumbent stone circles are probably distantly related).

We can only guess at the strange beliefs and eerie ceremonies that have left their mark in the recumbent stone circles. Burl speculates that "the rites enacted in the rings were closely connected with the flourishing and dying of plants, crops, animals and human beings in the short-lived world of four thousand years ago."[14] A concern with the moon's movements may have been inseparable from the anxiety of ensuring a good harvest or of pleasing an ancestral spirit.

DID THEY PREDICT ECLIPSES?

Was this remarkable awareness of the moon shared with other communities outside northeast Scotland? Gerald S. Hawkins began the search for lunar orientations when he published his original study of Stonehenge in 1963. His arguments for the existence of alignments to the long-term lunar cycle at Stonehenge are now accepted by many archaeologists, although the statistics connected with these lines remain controversial. The evidence of only a single site often presents difficulties to the investigator, no matter how carefully the research is carried out.

Thom avoided this problem when he proposed his first lunar theory in 1967, because the alignments were drawn from about forty different sites scattered all over Britain. At this stage, he suggested that the swing of the moon was observed with only moderate accuracy. However, Thom soon developed and refined his theory. By 1971, when his second book, *Megalithic Lunar Observatories*, appeared, Thom and his family had begun a unique effort to discover extremely accurate lunar alignments among the stones of Brittany as well as Britain. According to his theory, the moonwatching carried out by the megalith builders was astonishingly precise; Thom claimed that many of the sight lines he discovered were accurate to *at least three arc minutes* (one twentieth of one degree).

Why should prehistoric observers have troubled themselves with such fussy standards? Their motives, according to Thom, were governed not so much by religious compulsions as by intellectual curiosity. In their book published in 1978, Alexander Thom and his son Archie describe the megalithic astronomer as a "potential scientist. . . . He did not know where this was leading him any more than today's scientist really knows what the outcome of his work will be, but the earlier people were motivated by the same urge to study phenomena that drives the scientists of today."[15] The Thoms also believe that the lunar lines were necessary for an ·elaborate technique of predicting eclipses. If the erectors of the stones really had succeeded in predicting eclipses on a regular basis, then they might indeed deserve the title of "potential scientists." But how likely is it that the observers progressed that far?

We generally associate eclipses with the spectacular moment when the disc of the moon entirely masks the face of the sun, plunging the day into an eerie, unreal twilight. We can easily imagine the alarm that prehistoric people would feel at such a moment. However, total eclipses of the sun are rare events. The moon's shadow casts a narrow belt across the earth's surface only a little over one hundred miles wide, so the chances of an ordinary person being in the right place to see one are limited. While partial eclipses are more frequent, they often escape notice because of the intensity of the sun's glare and the difficulty of staring at it directly.

A total solar eclipse.

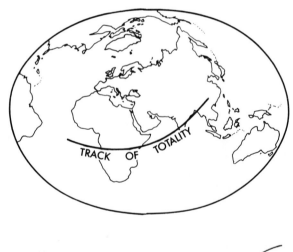

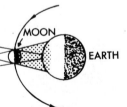

During a total solar eclipse, the moon's shadow covers only a small area of the earth's surface. Diagram, top, shows the narrow track along which the eclipse of February 16, 1980, was visible.

By comparison, eclipses of the moon are much more common and widely visible, although they are less dramatic. When the earth passes between the sun and the moon, the moon falls into the earth's shadow and is eclipsed. Because the earth's shadow is comparatively large in relation to the area of the moon, the darkness obscuring the lunar surface may persist for several hours, and it can usually be seen from at least half the earth. A single observer in one place is able to see an average of about one lunar eclipse each year.

Furthermore, the earth's atmosphere often bends the solar rays so that a striking ruddy glow accompanies the progress of the shadow across the moon. In 734 A.D., for example, the Anglo-Saxon Chronicle records the moon's ominous appearance "sprinkled with blood," an event that was taken to herald the death of the Venerable Bede. Similar thinking might have led the megalith builders to associate lunar eclipses with the fortunes of the living or with the spirits of their ancestors.

There will always be a full moon at the time of a lunar eclipse because it occurs when the sun and moon are opposite each other with the earth in between. But predicting which full moon will be obscured is not at all simple.

It is hard to visualize exactly how the orbits of the earth and moon interact with the fixed position of the sun in space. We can easily grasp the basic idea, though, that the occurrence of an eclipse is determined by three different motions: the orbit of the earth around the sun (one year), the orbit of the moon around

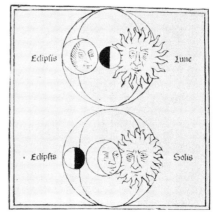

A 17th-century Venetian woodcut of eclipses.

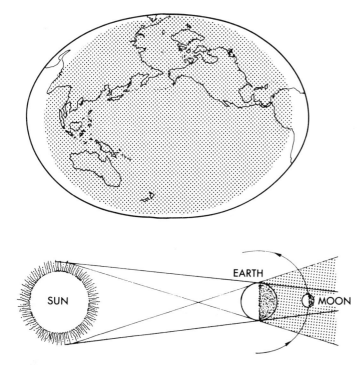

By contrast, a lunar eclipse like that of December 30–31, 1982, is visible from most of the earth.

the earth (one month), and the slower variation of the moon's orbit itself (the long-term cycle of 18.61 years). All three motions must coincide for an eclipse to be possible.

The result of these combined celestial movements is a set of complex and shifting patterns of eclipses over the years. Many different regular patterns would be apparent in a long list of accurately recorded eclipses, although some patterns would be better than others. "Better" here means that one cluster of predicted dates based on a given pattern will go on successfully forecasting these events with only a slight error over several centuries, while another will fall out of step with real eclipses and be useless after only a couple of repetitions.

These patterns are difficult enough to understand today, but they would have been even more inscrutable to the prehistoric skywatchers, who had neither our tradition of written mathematics nor tables of eclipse observations compiled from every corner of the globe. Was it possible for them to make predictions?

If we suppose that they were unconcerned with precision but simply wanted to establish rough warning periods when eclipses might occur, much simpler methods than Thom's technique were available. For instance, about half of all lunar eclipses are separated by intervals of six months. Assuming that this rhythm had been noticed, the observers would have been specially alert at every sixth full moon. If the moon rose shortly after the sun set, and if they were exactly opposite each other when it

The Qagyuhl of the Northwest coast of America believed that a lunar eclipse resulted from the attempt of a sky monster to swallow the moon. By dancing around a fire of smoldering clothing and hair, they hoped to compel the monster to sneeze and disgorge the moon.

happened, a lunar eclipse was bound to follow. Only simple observations would have been involved, and no use of numbers or calendars, nor any understanding of the lunar movements, would have been really necessary.

On the other hand, a demand for more accurate predictions would present many difficulties. During a single lifetime, many eclipses would be missed because of cloudy nights and because they were invisible from the latitude of the British Isles. If we suppose that an accurate calendar was in existence, the next question is whether a long sequence of dated eclipses could be memorized and passed down from one generation to the next.

Today we rely so much on the printed word that we underestimate the power of oral memory and its importance for storing up traditional knowledge among nonliterate peoples. For instance, the Druids of Britain and Gaul probably committed long verses of some sort to memory. According to Caesar, who drew on earlier sources dating from the late second century B.C., the Druids attended "schools" where, he claimed, some spent twenty years in training.

Nevertheless, it is hard to believe that eclipses really could be predicted from memorized lists of incomplete observations. Some more permanent recording device would surely have been essential. Did such aids as written numbers and dates exist at the time of Stonehenge?

Most perishable materials on which symbols could have been carved, painted, or woven have, of course, failed to survive to the present day. However, prehistoric rock carvings have been preserved in many parts of the British Isles. These carvings are the so-called cup-and-ring marks, which consist of circular motifs deeply chiseled on flat, natural outcrops of bedrock. The designs almost always take the form of a small ground-out hollow, or "cup," in the center, surrounded by one or several concentric rings. A single radial groove often cuts across the engraved rings.

Attempts to "decode" this baffling symbol, which appears in

A cup-and-ring motif at Roughting Linn, Northumberland, the most extensively decorated rock surface in England.

much the same form in Argyll as it does in Kerry, have been pursued by scholars and enthusiasts for over a century. Although the cup-and-ring *could* stand for the sun—among a host of other possible meanings—there is no convincing evidence that the carvings were star maps or eclipse records, or that they were even related at all to the sky. The range of different symbols is too narrow to represent a system of shorthand writing or arithmetic. Their meaning remains utterly mysterious.

It seems unlikely, then, that signs for numbers or dates had been invented. If so, they would surely have been commemorated on the rock surfaces alongside the cups and rings, and perhaps even on pottery and personal ornaments as well.

This is not to deny the likelihood that counting and numbers existed in prehistoric Britain. In particular regional groupings of stone circles, there was usually a preferred choice of how many stones were to be erected. Clearly there were local customs about numbers (the idea of lucky numbers, perhaps) that went hand in hand with other traditions about how a monument should be set out. Incidentally, these numbers rarely corresponded to cycles of the sun and moon. *

Simple tallying aids such as beads on leather thongs or pebbles kept in skin bags might well have been used to keep track of large numbers. But there must have been limits to this kind of numbering system. One is naturally skeptical about theories that call for higher mathematics from the Stonehenge people. Fred Hoyle, the renowned astrophysicist and science-fiction novelist, proposed the most complicated of all the theories about eclipse prediction at Stonehenge, demanding an ability to handle and record five-figure numbers, long division, and fractions. Although the Stonehenge people were as intelligent as ourselves, it is surely implausible that they would have developed such advanced arithmetical skills.

Thom's eclipse theory also places considerable demands on the ancient observers. His complex forecasting technique requires the detection of a tiny, regular "wobble" in the orbit of the moon, first noted during the late sixteenth century by the Danish astronomer Tycho Brahe. This small variation could not be seen each and every month, but only when the moon approached the end of its swing through the sky every 18½ years. Moreover, it consisted of a mere nine minutes of arc (less than one fifth of one degree).

The recognition of such a slight motion was a problem similar to knowing the exact day when the winter or summer solstice occurred. At sites such as Ballochroy and Kintraw, Thom claimed that the ancient observers had lined up the last flash of

Many of the megalithic observatories studied by the Thoms consist of isolated standing stones in Scotland. Above, a proposed lunar observing stone at Camus-an-Stacca, Argyll, and, above right, an alleged solar marker, Clach Mhic Lheoid, on the Island of Harris in the Outer Hebrides. It is always difficult to know whether these single stones really do indicate distant features on the horizon.

---

*The exception was Stonehenge, where a connection with the moon is, in fact, suggested by numbers such as thirty and nineteen, which are incorporated in its layout.

the sun with a distant mountain, allowing them to follow the tiny daily change in the sun's orbit at the time of the solstices. He now proposed that the observers stood where they could see one edge of the moon rising or setting against a far-off peak or dip in the hills. This type of alignment led them to recognize the "wobble" and devise their ingenious method of forecasting eclipses of the moon.

The proof presented by Thom deserves to be considered properly, but the lunar theory, which depends so much on high-precision measurements and observations, has aroused skepticism in several expert astronomers. They worry, for instance, about the effects of prolonged bad weather or horizon haze on work at the observatory. How could a thin crescent moon be lined up against a stone or the horizon? Such practical difficulties might well have prevented the observer from seeing any regular pattern in the subtle changes of the moon's movements every $18\frac{1}{2}$ years.

In his book *Megalithic Lunar Observatories,* Thom presents a selection of forty sight lines surveyed at twenty-three different standing-stone sites. Most of these sites are scattered in the Highlands of northern and western Scotland, where dramatic, jagged horizons are commonplace. Thom assumes that the ancient astronomers at each site lined up a distant mountain with the moon, even though the ruins actually present there often give little or no indication of the mountain involved. In fact, the monuments are frequently surrounded by so many dips and notches on the skyline that there is an obvious possibility of a potential moon alignment occurring quite by chance.

The situation is different in the gentle, undulating heathlands of southern Brittany. Here the relatively smooth horizons led Thom to investigate a different method of observing the moon. Since no mountains were available, he claims that a manmade target would have been erected, one so big that it could be seen from many miles away. This is how Thom attempts to explain one of the most spectacular engineering achievements—or, possibly, disasters—of the ancient world.

6. The Riddle of the Fairy Stone

THE GREAT BROKEN STONE OF BRITTANY

The remains of the largest of all the prehistoric European megaliths lie in the outskirts of the small fishing port of Locmariaquer in southern Brittany. Its Breton folk name, the Men Er Hroeg (the Fairy Stone) seems ironic, unless it reflects a sense of astonishment that human power could be responsible for shifting so huge a block. Most authorities figure the stone's original length to have been about sixty-five to seventy feet, and its breadth near the base to have been over thirteen feet. It probably weighed an awesome 260 metric tons.

The labor invested in the Fairy Stone can scarcely be imagined. Since the natural process of weathering creates detached blocks of granite in a chaotic variety of shapes and sizes, a preliminary quarrying operation was probably unnecessary. The source of the original outcrop, however, is uncertain. The stone is composed of a pale gray granite, polished to glassy smoothness by the passing of centuries, and flecked with white quartz crystals that catch the eye. This rock does not resemble the local bedrock, a dull calcareous granite. The most plausible explanation for this difference is that the source of the Fairy Stone was a local outcrop subsequently eroded away or submerged by the rising sea level of this region. In support of this idea, the Breton archaeologist Zacharie Le Rouzic noted that prehistoric tombs and even recent garden walls at Locmariaquer were composed of the same crystalline rock, and that traces of it could be found beside the sea.

The profile of the base is an attractive lozenge shape with rounded corners; a few surviving tool marks on the surface suggest that this shape was produced artificially. This would have been an unenviable task, involving endless repeated bashings with the aid of small stones.

Once it was finished, the great monolith had to be eased to an upright position. Presumably the builders used a ramp of earth or boulders, wooden levers, and ropes woven from vegetable fiber or hide (though none of these materials has survived). A comparison drawn from more recent times illustrates how formidable their task must have been. In 1586, Pope Sixtus X commanded

Bottom, *the broken remains of the Fairy Stone in the outskirts of Locmariaquer, southern Brittany, have intrigued and perplexed visitors for centuries. Profile view of the base of the Fairy Stone, below, illustrating the lozenge shape that was the result of artificial workmanship.*

the erection of an Egyptian obelisk in St Peter's Square in Rome. This monument was longer than the Fairy Stone, but, according to one estimate, weighed twenty tons less. Even with a work force of 850 men, seventy horses, and forty-six cranes, the operation was nearly a disaster. The obelisk was caught at a precarious angle as the ropes raising it jammed in their blocks. A sailor is supposed to have shouted "Water on the ropes!"; this action enabled the obelisk to be maneuvered to a safe, upright position.

Did a similar catastrophe strike at the Fairy Stone several thousand years ago? Today, instead of a proudly erect marker dominating the coastline for many miles, only its broken remains lie prone like a great beached whale, close to the cottages and vegetable gardens of Locmariaquer.

The most widely accepted version of events is that lightning or an earth tremor toppled the stone so that it fell and smashed into four pieces (a fifth piece near the top may once also have existed). Mild seismic shocks are quite common in this region, but is this adequate to explain the curious position in which the fragments came to rest? The top three pieces are in a line roughly from south to north, while the base lies almost opposite to this, pointing northeast. In other words, the two major sections of the stone appear to have been flung apart from one another with considerable violence. Was an earth tremor responsible for the rocking or twisting motion that seems to have been applied to the stone?

This explanation is popular, and it may well be correct (the close spacing of the fragments supports it), yet there are other possibilities. A Californian physicist, Robert D. Hill, speculates that the stone fell and broke by accident while it was being hauled upright, and that only the stump was eventually reerected by the original builders. In turn this piece, too, crashed down in the other direction—or perhaps it was deliberately pushed over at a later period. Certainly neither the entire stone nor its stump fell in recent history, since navigational logs dating back to 1483 mention other smaller prehistoric monuments as landmarks visible from the sea, but not the Fairy Stone.

Alexander and Archie Thom believe that the ancient Bretons did succeed in erecting the stone. They suggest that the reason for its immense size and bulk was scientific necessity; it had to be

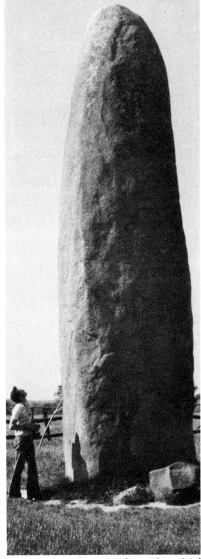

The standing stone of Dol near Saint-Malo in northern Brittany is 31 feet high, weighs at least 125 tons and was probably transported to the site from an outcrop over 2 1/2 miles away.

Two nineteenth-century antiquarians debate the demise of the Fairy Stone.

huge and tall to be seen against the disc of the moon. The observers would line up the edge of the moon with the distant stone, just as others had done in Scotland with far-off mountains.

There is a total of eight main directions for the rising and setting points of the moon. By tracing these directions as lines across the map from their center at the Fairy Stone, the Thoms hoped to discover somewhere along them the remains of sites where the observers had stationed themselves to watch the moon. If prehistoric stones were found at exactly the right spot—at the correct angle for the moon and at a place where the Fairy Stone would have been visible in the distance—then the lunar theory would surely be convincing.

Thom and his team patiently explored the terrain in the required directions. Along four of the eight sight lines, no ancient ruins turned up. When they inspected the four lines for the rising moon, however, the Thoms reported a total of seven different locations where they had found promising stones.

In 1980 I decided to visit these sites to record what could be seen there. At the back of my mind was the question of whether the role of chance had been properly assessed in the case for the observatory. After all, if we ruled *any* line across a map of the region with the thickest concentration of standing stones in Europe, would it be surprising *not* to pick up several sites along the way? More observing places were found along the lunar line that passed close to the great cluster of monuments at Carnac than in any other direction. In other words, the most sites had

Many of the Thoms' proposed lunar observing sites at Carnac involve insignificant-looking remains. At Kerlagad, the unimpressive stone A, opposite top, is claimed to be a lunar marker, while the magnificent stone G in the nearby field, opposite below, apparently has no astronomical significance.

Map of the Carnac region of southern Brittany, showing lunar observing sites for the Fairy Stone identified by the Thoms.

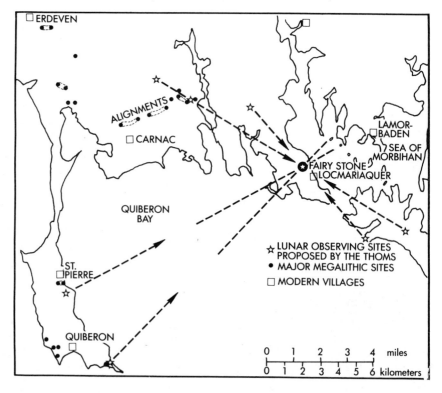

been found where the most stones existed. This was a hint that the lunar observatory might be simply the accidental product of a landscape rich in prehistoric ruins.

Furthermore, the Thoms' concern with accuracy led them to consider other values besides the eight basic angles for the moon. (These corresponded to the effects of the lunar "wobble" and other astronomical factors). The result was that a total of thirty-two possible sight lines to the moon was available. This extra ration of lunar lines increased the likelihood that the Thoms would find sites which fitted their theory. Obviously, it was tricky to evaluate the part played by chance in the whole affair.

One way of doing so was to see how consistent the possible observing sites appeared to be. Did the stones marking them look similar in size and layout, suggesting that they had all been set out together with the same purpose in mind, to study the moon? Were the stones reasonably impressive? It seems improbable that the ancient skywatchers who expended so much time and effort observing the moon and procuring the mighty Fairy Stone would erect only insignificant little markers at the crucial positions where the watch on the moon was to be kept. How did the stones identified by the Thoms match up to these expectations?

The stones are undeniably a "mixed bag." They range dramatically in size from one magnificent six-meter-tall giant to several pygmy specimens less than a meter high. In some cases, the shape of a stone directs the eye in roughly the same direction as the astronomical alignment, but in other cases the positioning of the flat sides seems to point at quite a different angle.

Most important of all, the sites represent more than one kind of monument. Some stones stand in peaceful isolation among the pine woods. Others are found beside or on top of long burial mounds. One stone was close to another kind of burial structure, a stone-chambered passage grave. Another site took the form of stone rows. In several cases I was doubtful as to whether or not the remains were truly ancient (including one "fallen slab" that appeared to be an exposed lump of bedrock). As I visited each site reported by Thom and his team, my suspicions grew that the Fairy Stone was a lunar observatory by accident rather than design.

Such a conclusion does not mean that the Thoms' work is useless, nor does it rule out the idea that the stones of Brittany were connected in some way with astronomy. In fact, all those interested in the Carnac monuments have reason to be grateful for the Thoms' extensive surveying efforts in the area. Their labors might have been even more rewarding if they had tested more than just one narrow type of astronomical theory. Did imprecise symbolic alignments exist, comparable to those of Newgrange or the Scottish recumbent stone circles? The possibility that the celebrated avenues of standing stones at Carnac and

COWS, STONES, AND SEX: THE MEANING OF THE BRETON MEGALITHS

some of the great stone tombs nearby were *roughly* oriented to the sun or moon deserves a thorough investigation.

In any case, the evidence of archaeology and folklore does suggest that beliefs about the sun figured in ancient Breton traditions. The summer solstice at Carnac, for instance, was once a grand festivity. Young people used to collect firewood and build an immense pyre on top of a huge burial mound in the outskirts of town. James Miln, an English archaeologist, watched the preparations during the 1870s as *tan heol*—"the fire of the sun"—was kindled on midsummer evening:

> Just as the sun sets long clouds of smoke are seen rolling along the ground from the neighborhood of adjoining farms. If you ask the cause of all this you are told that these are fires prepared expressly for the cattle, with green broom and fern, so as to produce a dense smoke. In fact it is at this time that the cattle are brought home, and before being housed for the nighttime made to pass across these fires, the doing of which is supposed to preserve them from all maladies to come.[16]

Other traces of a cattle cult at Carnac persisted until recently. The annual fête, or *pardon*, held in September, was attended by pilgrims from all over Brittany. At its climax, cattle were driven solemnly up to the main doorway of the town church to be blessed by the local clergy. The discoveries of burned cattle bones and horns, often buried in considerable quantities under the huge

The seventeenth-century church front at Carnac displays a statue of the town's patron saint, Saint Cornely, beside two paintings of bulls surrounded by megaliths, said to be the work of a local shoemaker. Prehistoric carvings in the tomb of Mané-Lud, below right, may represent cattle horns.

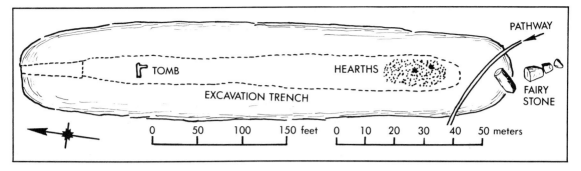

PATHWAY

TOMB   HEARTHS

EXCAVATION TRENCH

FAIRY STONE

0   50   100   150 feet   0   10   20   30   40   50 meters

Neolithic burial mounds near Carnac, suggest that such practices may have had their beginnings in remote prehistory.

Today few visitors realize that the broken fragments of the Fairy Stone are associated with a similar burial monument, now nearly obliterated by a modern car park. The enormous cigar-shaped mound, which probably dates to well before 3000 B.C., extends from the foot of the Fairy Stone in a north-northwest direction for a distance of over five hundred feet. Originally the great mass of the mound, about ninety feet wide and thirty feet high, must have been nearly as imposing as the gigantic stone at its southern end. The sides were reinforced with a wall of substantial stone blocks. The top was leveled off, and on this lofty platform, paved with pebbles, hearths once blazed fiercely. Cattle or horse leg bones littered the fireplaces.

Underneath the long mass of earth and rubble, a stone-built chambered tomb lay sealed off from the outside world. Inside this vault were traces of human bone, flint, pottery, fine axes crafted from semiprecious jadeite, and necklaces and a pendant fashioned from *callaïs*, an attractive local blue-colored rock. At a number of other long mound sites in southern Brittany, tall standing stones also presided over similar rich offerings of ashes, flint, pottery, and axes. These examples suggest that the erection of the Fairy Stone was planned at the same time as the raising of the huge long mound. In the builders' minds, the stone was probably associated with human death, the kindling of fires, and the act of offering up both beautiful necklaces and drab pieces of pot. Perhaps the stone represented a guardian "ancestor spirit," watching over the bones of the dead.

The Fairy Stone may have signified fertility as well as death. Some standing stones are so suggestively phallic in appearance that a sexual meaning seems obvious. It was certainly obvious to Breton peasants up to a century ago. Uncomfortable though it must have been, the local inhabitants of Locmariaquer had a custom of gathering around the Fairy Stone on the night of the first of May and sliding down the fragments with bare bottoms. Embracing prehistoric stones, or rubbing up against them, was regarded as a remedy for infertility or infidelity in many parts of Brittany. During the last century, young married couples, stark naked in the moonlight, would chase each other around one of the fallen stones in the Carnac avenues while their parents

Diagram of the Fairy Stone and the long burial mound associated with it, based on the plan of its excavator, Le Rouzic, published in 1909.

obligingly kept a lookout from the top of a stone located at a decent distance. (The custom was said to be a cure for infertility.)

The sexual aspect of the stones was regarded in a broad sense, including the health and well-being of beasts as well as humans. Fragments of standing stones were chipped off and boiled in a pan to make a curious broth, said to cure the sicknesses of cattle as well as people. Such beliefs may well have figured in prehistoric times, expressed on a grand scale in the ceremonies accompanying the erection of stones or the burials inside the tombs.

In seeking the purpose of the stones, it is, of course, unscientific to be swept away by a romantic twilight of folk beliefs, some of which may be only hundreds rather than thousands of years old. But it is also unscientific to disregard all forms of evidence other than the calculations of astronomy and geometry based on the outward appearance of sites today. The result, in the case of Professor Thom's research, is a kind of science fiction in which the prehistoric Bretons operate their lunar observatory with a passion for precision suspiciously like that of modern engineers and astronomers. While the megaliths were indeed connected with beliefs about the sun and moon, it is clear that these were only aspects of a complicated mass of ideas and practices involving the afterlife, ancestral spirits, and the fertility of beasts and humans.

We must be grateful to Thom and his colleagues for awakening interest in the intellectual skills of prehistoric Europeans. Yet many of the claims for a Stone Age science are now proving to be little more than an unconscious projection of our contemporary

technical world onto the silent ruins of four thousand years ago. There is little "hard" evidence to support the theory that accurate eclipse predictions were undertaken by the megalith builders, who would have found their task extraordinarily difficult in the absence of written recording aids. The lunar alignments are more convincingly explained as pious observances of a "sacred" rhythm, the $18\frac{1}{2}$-year swing of the moon. In the case of Scottish monuments such as the kerb-cairns and recumbent stone circles, the lunar cycle seems to have been linked to funerary customs and ritual gathering places. Since so many gaps remain in our knowledge of prehistoric Europe, it is unlikely that we will ever fully understand the motivations of the megalithic skywatchers. But they were surely driven by religious faith or superstitious fear, rather than the impulse of scientific curiosity so vital to the astronomer today.

Midsummer sunset at Hovenweep Castle, Utah, a remote and mysterious ruin located in the Four Corners region of the American Southwest. A room in the corner block at the right seems to have been used to observe the sun more than 700 years ago.

PART III

The Sky Priests of North America

7. The Hunters and the Heavens

THE ICE AGE MOONWATCHERS

Who were the earliest skywatchers?

The first human awareness of the heavens lies in the remote past, stretching back far beyond the invention of farming. The rhythm of the sun was, of course, connected with seasonal changes in vegetation and the activities of animals. An awareness of these changes must have been part of the consciousness of hunters almost as far back as the earliest scientific evidence of human remains.

Today there is little doubt that the major center of human evolution was the open grasslands and lake shores of East Africa, the setting for a number of dramatic anthropological discoveries during the early 1970s. For example, there was the surprisingly advanced skull known as "1470" unearthed by Richard Leakey, son of the famous husband-and-wife team Louis and Mary, and the relatively complete skeletal remains of "Lucy" assembled by Donald C. Johanson of the Cleveland Museum of Natural History. Such finds showed that the appearance of human characteristics dated back to at least three million years. By the time of the oldest known structure—a semicircle of large stones erected as a crude hut or windbreak some 1.8 million years ago at Olduvai Gorge in Tanzania—it is certain that early hunters were already engaged in transporting food and materials back to a home base. But the hunters were confined to their African homeland until a later period, when the control of fire became widespread, enabling them to settle and survive in much less hospitable regions. Early man's first tentative appearance in Europe seems to have been in the depths of the ice age known to European scholars as the Mindel and to Americans as the Kansan, some half a million years ago.

One of the earliest and most interesting prehistoric sites in Europe was discovered in 1965 by the French prehistorian Henry de Lumley at Nice, on the Riviera. The remains had not washed up on the shore, but were dug out of a beach nearly one hundred feet higher and over three hundred thousand years older than the present one. The most remarkable finds at this site, known as Terra Amata, came from the foundations of huts built of light branches, which were heated and paved inside with stones and skins. Areas free of debris near the hearths indicated where the occupants had curled up for sleep and warmth. The sand had preserved the imprint of a wooden bowl and a human footprint. Because of the absence of skeletal remains, we have little idea of what these ancient Europeans looked like, but evidence of their simple stone tools and their diet did survive.

The animal bones recovered from the site are mainly of young deer, supplemented by elephant, pig, turtle, fish, and oysters. Plant pollen preserved in the fossilized human feces tells us that

Artist's impression of the huts excavated by Henry de Lumley at Terra Amata, dating to about 300,000 B.C. The walls, built of light branches, were so insubstantial that the occupants had to protect their hearths inside with low banks of pebbles.

the meals were consumed during the late spring or early summer. Henry de Lumley believes that repeated visits to the site were made at this time of year, since one of the huts was rebuilt eleven times on the same spot. It appears as if the inhabitants of Terra Amata were pursuing a regular pattern of seminomadic life typical of hunter-gatherers in recent times, and were also recognizing seasonal changes in their surroundings. They had a "calendar," however crudely or unconsciously realized it may have been.

The next obvious step was to keep a simple tally of the passage of time, perhaps by compiling a sequence of marks for the days in each month. Several lumps of red ocher, a natural mineral pigment, were found inside the huts at Terra Amata, but their use at the site is unknown. There was certainly no regular practice of making marks on bone or stone for many hundreds of generations after the beach expeditions of the Terra Amata people. In fact, the widespread appearance of decorated objects, sculptured figures, and engravings and paintings inside caves did not occur until the later part of the last ice age (known as the Würm to European scholars and as the Wisconsin to Americans).

This was an age of relatively rapid change compared to the millennia that had gone before. Now, from about 40,000 B.C. onwards, men and women of modern physical build were everywhere present in Europe. To protect themselves from the winter extremes of a bitter glacial climate, the hunting bands of western Europe occupied the entrances of caves and rock overhangs. On the bleak plains of central and eastern Europe, where such natural shelters were unavailable, they built substantial houses of mammoth bones covered with skins. The hunters were tracking herd animals such as reindeer, wild horse, bison, and mammoth; in their annual migrations, these animals frequently drew the human groups many hundreds of miles from their seasonally occupied home bases. The Ice Age people possessed a remarkably refined stone and bone tool technology, and they left evidence of their highly developed artistic sense, as the vivid paintings at Lascaux, Altamira, and other caves remind us.

In 1963, Alexander Marshack, who was then a science writer researching a book on space travel, examined a photograph of an ancient bone with distinct groups of lines engraved on its surface. It struck him that such marks might be purposeful and not merely the product of casual scratchings or decoration. The idea led Marshack to examine engraved Ice Age objects in the collections of European museums. Many of these objects displayed rows of small lines and notches that had either escaped the attention of previous scholars or had simply left them puzzled. The existence of engraved sets of lines, or "notations," was in fact discovered more than a century ago. Antiquarians and scholars referred to them as *marques de chasse* ("hunting marks"), a term which suggests the improbable idea that an ancient hunter needed to keep a score of the animals he had brought down.

A far more promising theory emerged from Marshack's study of the carved bones. Over several years he developed a technique of examining them with the aid of a low-power zoom microscope and photographic enlargements. This method revealed that many sets of markings were made not just with one single stone tool, but with several. The different tool points left distinctively shaped grooves in the bone. The obvious conclusion was that the marks were not all created on a single occasion but were accumulated in a sequence over a period of time. Such a sequence presumably represented a record of some kind, and so could have been used to keep track of a hunter's kills; but was there a more plausible explanation?

One obvious possibility is that the Ice Age engravers were compiling a rudimentary calendar by observing and noting the phases of the moon. No obvious lunar designs or symbols are visible on the bones. However, if the theory is right, the sequences of marks carved by different tool points can be expected to fall into a rhythm reflecting the changes in the moon's appearance. In 1970 Marshack published a detailed examination of six carved objects which seemed to confirm the idea that they

An Ice Age bâton *of uncertain use from* Le Placard, Charente, France, *exhibiting a typical sequence of carved notches or "notations" on its surface. Note the fox head fashioned at one end of this antler implement, which measures a little over a foot long.*

were lunar tallies. In subsequent papers and his book *The Roots of Civilization,* Marshack presented brilliant photographs and descriptions of other engravings that also seemed to have a connection with the moon.

It was difficult, of course, to *prove* the connection, because with observations of this kind, the number of days in each month or part of a month would vary. Sometimes a month would appear to be twenty-eight days long, another time thirty-one; these counts would also depend on just how the observer treated the period of invisibility around the time of the new moon. Furthermore, Marshack considered it likely that half and quarter months were recorded, and so numbers like thirteen, five, and eight also became significant in his studies of the markings. The bigger the choice of possible lunar numbers, the more likely it is that a particular sequence of marks will fit the theory. Uncertainties of this kind mean that Marshack's work is not really provable. On the other hand, his lunar theory remains the most sensible explanation of certain Ice Age engravings.

On some objects, the carved signs accumulated over a period of time consist not merely of simple lines or notches, but of more complex patterns including zigzags and lattices, and there are even little symbols that seem to represent animal heads or plants. These sequences may well represent the matching of a lunar count against seasonal changes in the hunter's surroundings, such as the arrival of migrating salmon or the first appearance of spring buds. A keen awareness of seasonal events is also suggested by many of the beautiful artistic depictions of animals on cave walls and on engraved objects. For instance, in his observations on a bone from the site of Montgaudier in the Charente district of western France, Marshack drew attention to the tiny realistic carving of a male salmon. It is shown with a slightly hook-shaped lower jaw, a characteristic that develops only during the spawning season. Perhaps the famous animal paintings of Lascaux and Altamira were inspired by sacred seasonal stories, myths, or ceremonies. The meaning of the animal images may be richer and more complicated than many popular theories suggest (such as the common idea that they were painted as part of "hunting magic" rites).

It is obviously impossible to do justice to Marshack's work in a brief summary. Perhaps his most vital achievement was to show that the earliest attempts to record the passage of time began

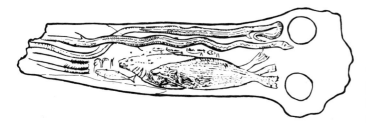

The design engraved on the surface of an Ice Age object from Montgaudier, Charente, rolled out as a flat drawing for clarity.

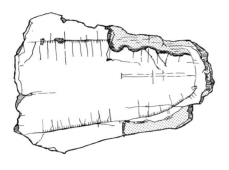

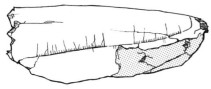

Engraved mammoth tusk from Gontzi in Ukraine.

Opposite, a sample of early prehistoric hunting-and-gathering sites, documenting the slow colonization of the New World from Asia. In most cases the dates shown are extremely rough, and several are controversial.

during the last ice age, in the period from about 35,000 to 9,000 B.C. The use of actual numbers and counting by the original engravers may not have been essential to the accumulation of the carved notches. But, as Marshack points out, even though the engravings do not constitute formal writing or arithmetic, they nevertheless represent an essential step toward the invention of such a system. The urge to understand and create order in the changing natural rhythms visible on earth and in the sky had its origins in the world of Ice Age hunting, not in its postglacial aftermath. The complexity of hunting-and-gathering life was sufficient to arouse intellectual activity in people who, by 35,000 B.C., were anatomically identical to ourselves. The invention of farming and the spread of a settled existence were merely further steps along the road of human inventiveness that had begun tens of thousands of years earlier.

The carved bone artifacts of the Ice Age are not found exclusively in areas of western Europe where cave paintings have survived, such as southwest France and northern Spain. Indeed, Marshack has shown that the practice of accumulating carved lines and images on a bone object was widespread in central and eastern Europe. The designs engraved on the bones (some of which Marshack thinks may be fish and water symbols) differ from those in the west, but the basic act of adding one motif after another over a period of time seems identical.

Some of the most interesting carvings come from the valleys of the Russian Ukraine. The engraved objects were excavated from the ruins of settlements ingeniously constructed from mammoth bones and hides because of the shortage of timber on the surrounding plains. At the sites of Gontzi and Adveevo, the occupants of this type of dwelling left behind bones with elaborate linear sequences carved on them. The original of an engraved ivory tusk from Gontzi, housed in a Russian museum before World War II, is now unfortunately lost, but from a drawing Marshack was able to advance a tentative interpretation of the markings as a record of the lunar phases.

The discoveries of possible lunar tallies apparently extend to the easternmost extremities of the Soviet Union. In 1978, for instance, the Russian anthropologist N. N. Dikov reported the finding of "a kind of moon calendar" from the sunken floor of a house that was probably built around 10,000 B.C. The site was near Lake Ushki in the central Kamchatka peninsula, on the eastern fringe of Siberia, an area of particular iterest to prehistorians as they attempt to trace the earliest immigrants to the New World.

WHO WERE THE FIRST AMERICANS?

Today there is no serious doubt that the Americas were first settled by Asian hunter-gatherers who entered the continent by crossing the Bering Strait. A wealth of biological and archaeological evidence supports the case. For example, one of the most convincing indicators of the ancestry of the American Indian is

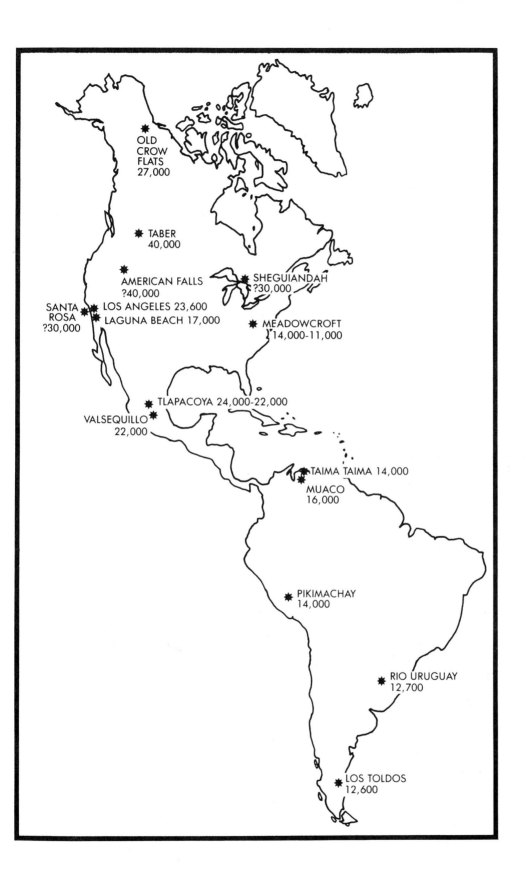

OLD
CROW
FLATS
27,000

TABER
40,000

AMERICAN FALLS
?40,000

SHEGUIANDAH
?30,000

SANTA
ROSA
?30,000

LOS ANGELES 23,600

LAGUNA BEACH 17,000

MEADOWCROFT
14,000-11,000

TLAPACOYA 24,000-22,000

VALSEQUILLO
22,000

TAIMA TAIMA 14,000

MUACO
16,000

PIKIMACHAY
14,000

RIO URUGUAY
12,700

LOS TOLDOS
12,600

the scooped-out appearance of their front incisor teeth, a highly characteristic Asian trait. With firm clues of this kind available, the controversies are concerned not with the question of origins but with the question of the exact time the colonization began.

The existence of seagoing craft during the Ice Age is now widely accepted, since we know that Australia was settled by at least 30,000 B.C. The pioneers were not therefore necessarily confined to the periods when the Bering region consisted of a continuous land bridge between the two continents.

The numbers of the early immigrants must have been very low, since skeletal remains in particular are few and far between. Controversial human finds from California and Alberta may go back to over 40,000 years ago, but the dates have not been clearly established. There is little doubt that a refined tradition of crafting antler and bone tools was present in the Yukon basin by 30,000 B.C. Many other sites throughout the Americas have yielded radiocarbon dates that enable us to trace the gradual spread of human settlement. By about twelve thousand years ago, the process was complete, for by then even the far south of Argentina was occupied. When all the dates are plotted on a map, the colonization of the Americas takes on the appearance of a swift military campaign, whereas in reality it took at least fifteen thousand years to complete. This was such a long period of time that the hunters themselves can scarcely have been aware of their gradual expansion across the new land.

A moonlit migration of reindeer herders in the far north—a romantic engraving published in 1827.

Whatever the exact chronology involved, early hunter-gatherers in the New World brought with them cultural and technological practices that allowed them to flourish as successfully as

they had done on the windswept Russian plains. This is not to say that prehistoric stone and bone industries were identical to those in Siberia, for the special conditions of life in the new continent quickly encouraged technical innovations. It is a reasonable guess, though, that at an early stage North American hunters would have possessed the capability of observing and recording seasonal changes or lunar phases as their counterparts in Russia seem to have done.

Are there any signs of the stone or bone counting devices which they might have used to keep track of the moon? On remote rock outcrops in the Great Basin, stretching from southern Oregon all the way to the Mexican frontier, engraved patterns of dots and lines occur in large numbers, some of which certainly have the appearance of counting records. An unusually elaborate carved rock at Presa de La Mula, Nuevo Léon, on the southeastern edge of the Great Basin, features over two hundred strokes, subdivided into twenty-four separate units. In 1982, W. Breen Murray, an anthropologist at the University of Monterrey, interpreted this engraving as a calendar covering a period of seven lunar months. There is no way of dating this and similar examples of rock art with any certainty, but Murray argues that they were the work of hunter-gatherers in the region, possibly several thousand years ago.

Elsewhere in the Great Basin, in central and southern Nevada, a hunting-and-gathering existence was pursued in a relatively unchanging environment for at least eight thousand years until modern times. During the 1970s, remarkable engraved stones were unearthed at several sites in this area. Archaeologists have dated some of these small slabs almost as far back as the beginning of this period. Many are quite portable and show signs of constant handling.

At the Gatecliff rock shelter located in the desolate juniper-and-piñon-covered hills of central Nevada, nearly 450 engraved limestone slabs have been found since excavations began in 1970. The designs vary from crude scratches to regular geometric patterns. Trudy Thomas, a student in the Art History Department at Columbia University, has examined the slabs from the site, with the help of technical advice from Marshack. Her preliminary report concludes that many of the designs on a single slab were accumulated over a considerable period of time, in a manner comparable to the earlier Ice Age engravings in Europe.

While it will always be difficult to be sure whether a particular carving represents a lunar tally, the Great Basin discoveries do suggest that hunter-gatherers in the New World had the capacity to make abstract "notations" many thousands of years ago. In the absence of hard evidence, we cannot be sure exactly when observations of a seasonal or astronomical nature were first recorded in North America. But they probably occurred long before the rise of farming, stimulated by the demands and opportunities of the hunting-and-gathering way of life.

THE SHAMAN'S SKY VISIONS

The minds of early settlers in the Americas cannot have been occupied solely with practical thoughts of subsistence and survival. They must surely have had special ideas about the universe, mythical explanations of the world that were rooted in the experience of generation after generation of hunter-gatherers. Many scholars have detected a connection between creation myths and stories about the sky in widely scattered regions. Certain fundamental beliefs, such as a preoccupation with the sacred four directions of space, were shared by groups as diverse as the highland Maya of Mexico and the nomadic hunters of the Great Plains. While it is only guesswork, one is tempted to imagine that basic elements of the traditions of all Native Americans stem from a very remote past, back to their common background as prehistoric hunter-gatherers.

Above and opposite, *typical engraved stones from the Gatecliff rock shelter in central Nevada.*

Some of the most striking parallels linking North, South, and Central America emerge from the study of shamanism. This was never a formally organized religion, but, rather, a complex web of ideas and rites centering on the individual practitioner, the shaman himself. He was the agent through whom contact might be established with the supernatural world. This contact took the form of ecstatic visionary experiences, often induced by hypnotic chanting or drumming, and sometimes by hallucinogenic substances. The objective of these rites was usually to cure the sick, or to exert influence over the weather or the behavior of hunted animals. A shaman might be initiated through dreams or visions sent by guardian animal spirits, who thereafter became his helpers in acts of healing or divination.

The study of shamanism provokes disagreement among scholars. A particularly difficult question is whether it represents a primitive type of religious experience common to all hunters everywhere, or whether it has unique historical and geographical roots. For instance, there are strong similarities between shamanism in North America and in Asia, which might lead us to suppose that these practices go all the way back to the Ice Age. But some of the similarities may be due to the fact that Eskimo hunters and Siberian herders shared the same general kind of environment. The characteristic trance of the shaman, for example, may be a psychological state brought on by desolate and solitary surroundings, a kind of "cabin fever" or "Arctic hysteria."

Another possibility is that mythological stories were spread slowly by contacts between the Asian and American continents that continued long after man first set foot in the New World. Such specific beliefs as the power of "thunderbirds" and the existence of the "world tree," with its roots in the underworld and its branches in the sky, are typical of cosmologies in the northern latitudes. They may have passed by word of mouth among traders and shamans over a long period of time.

Some of the correspondences between widely distributed ideas about the sky are quite remarkable. Recent work by a scholar of pre-Columbian literature, Gordon Brotherston of the University of Essex, England, demonstrates this very clearly. The soul journey was a visionary exercise performed in a trancelike state by the shaman in order to awaken his powers. Brotherston notes how the essential features of the soul journey coincide in sources as diverse as Mayan manuscripts from Mexico and Algonquin birchbark rolls from the northeastern United States.

In certain Mexican creation stories and epic tales, the soul of a deceased hero is described as passing to the east, where it

A Siberian shaman performing an act of divination, or fortune-telling, photographed in the 1890s among the reindeer-herding Chuckchi. The subject's head is bound with string and tied to the shaman's stick. If the answer is favorable, the subject's head is believed to rise of its own accord. If not, the shaman will be unable to lift it from the ground.

The special sanctity of the four cardinal directions is a fundamental belief of most Native Americans. The benevolent Mountain Spirits of the Mescalero Apache are thought to live inside mountains in eastern New Mexico. During dramatic firelit ceremonies, dancers impersonate the Spirits, each representing a cardinal direction, together with an attendant clown. The rituals continue today; this photo dates from about 1889.

descends into the underworld. In these texts, the visit to the land of the dead is identified with the period of eight days during which Venus is invisible between its appearances as the Evening Star and the Morning Star. Rising from the underworld, the resurrected hero then climbs up the steps of the sky. Finally he is united with the creator in the "heart of heaven," the zenith or noonday position of the sun located vertically over the earth.

In many North American sources, the progress of the soul after death and the ecstatic journey of the shaman were envisaged in almost identical terms. In the following account, a liturgy collected around 1910 among the Winnebago Sioux of Wisconsin, God, or "Earthmaker," awaits the arrival of the soul

> . . . in great expectation. There is the door to the setting sun. On your way stands the lodge of Herecgunina [the underworld] and his fire. Those who have come [the souls of brave men] from the land of the souls to take you back will touch you. There the road will branch off towards your right and you will see the footprints of the day on the blue sky before you. These footprints represent the footprints of those who have passed into life again. Step into the places where they have stepped and plant your feet into their footprints, but be careful you do not miss any. . . .[17]

Accompanied by the spirits of dead warriors, the soul arrives at the Earthmaker's Lodge at the zenith overhead, where sacrificial offerings are deposited. Earthmaker then addresses the soul as follows:

> "All that your grandmother has told you is true. Your relatives are waiting for you in great expectation. Your home is waiting for you. Its door will be facing the mid-day sun. Here you will find your relatives gathered."[18]

Symbol from an Algonquin bark scroll depicts the shaman's soul "walking up the sky" from the east to the zenith.

The four directions shaped many aspects of religion in ancient Mexico. In this page from a central Mexican manuscript, the Codex Fejervary, mythological figures express a complex scheme of opposed and balancing forces, all contained within a sacred cross-shaped framework of calendar units and the four directions of space. The manuscript dates from the 15th century A.D.

It is far from certain why correspondences of this type exist between the celestial mythology of ancient Mexico and that of the Great Plains. One explanation is that the influence of the great civilizations of Mexico was transmitted through trading contacts for several centuries before the arrival of the Spaniards. From these contacts, Mexican religious ideas may have passed indirectly into the cosmologies of North America.

Another possibility is that the common heritage of hunting and gathering shared by all Americans in remote prehistory helped to shape their unity of belief and outlook. In the thousands of years during which they led a seminomadic existence, we can probably trace not only the origins of rudimentary astronomical observations and calendar keeping, but also the rise of distinctive Native American attitudes about the sky and about the natural forces around them.

Were these unusual, inspired individuals, the shamans, responsible for passing on the earliest American sky lore? It seems probable, although without hard evidence, the existence of such figures in ancient prehistory must remain only a speculation. Certainly, by the time anthropologists began to study the North American hunting tribes in the late nineteenth and early twentieth centuries, they were steeped in elaborate cosmological ideas, many of them inspired by shamanism. Theirs was an outlook strikingly different from our own view of the universe. In particular, the anthropological accounts reviewed in the next chapter reveal that in many Native American communities, there existed until recently an attitude toward time that was quite out of tune with the precision and urgency so characteristic of our own culture.

Cosmic symbols on a "medicine shirt" worn by an Apache shaman.

8. The Counting of the Moons

How did early calendars develop from the crude scratches of the prehistoric moon counts? Did the ancient skywatchers share our passion for the logical ordering of time, and did this drive them on to make more exact measurements of the passing hours and seasons?

Our own dependence on accurate time information is so strong that the "practical" need for a precise timescale in any society may seem obvious. Our addiction to clocks and watches makes it difficult for us to imagine a less systematic attitude toward the past and future. Often we visualize time as a stream continuously slipping away from us. It can also seem like a substance to be measured out in definite quantities and matched carefully against the pattern of our everyday activities. The obsession of our culture with this short-term ordering of events is perhaps one reason why so many people find it hard to think more than a year or two ahead. The perspective of other cultures is different, as any visitor to Latin America knows. In his essay *Mornings in Mexico*, D. H. Lawrence observes that

> . . . to a Mexican, and an Indian, time is a vague, foggy reality. . . . Mañana, to the native, may mean to-morrow, three days hence, six months hence, and never. There are no fixed points in life, save birth, death and the *fiestas*. And the priests fix the *fiestas*. From time immemorial priests have fixed the *fiestas*, the festivals of the gods, and men have had no more to do with time. What should men have to do with time?[19]

The fact that our own outlook on time is a special one and not in any way "natural" or "instinctive" may seem an obvious point, but it is important when we interpret the evidence of ancient astronomy. The practical aspects of timekeeping and calendar regulation that seem so basic to us were not uppermost in the minds of many Native American people. To them, the watching of the sun and moon was often undertaken for religious reasons that are hard for us to imagine in all their strangeness and complexity.

Among the communities of the Southwest, punctuality in the sense that we understand it has little importance. From the notes of anthropologists who studied the Hopi tribe early in the present century, we know that the hour of certain ceremonies was timed accurately by such devices as the shadows cast on a wall, but the staging of other rites was erratic. The public dances that still impress visitors to the native southwestern villages, or pueblos, today, are often preceded by many hours of elaborate preparations in the secret underground ritual chambers known as *kivas*. During the 1950s, the anthropologist Edward T. Hall described his frustration at waiting for such a dance to begin in a pueblo near the Rio Grande during a freezing winter night:

In the church where the dance was to take place a few white townsfolk were huddled together on a balcony, groping for some clue which would suggest how much longer they were going to suffer. "Last year I heard they started at ten o'clock." "They can't start until the priest comes." "There is no way of telling when they will start." All this punctuated by chattering teeth and the stamping of feet to keep up circulation. . . . Those of us who have learned now know that the dance doesn't start at a particular time. It is geared to no schedule. It starts when "things" are ready![20]

Helmut Gipper, a German visitor to the Hopi in 1967 and 1969, encountered the same attitude:

Nobody could tell me at what time these dances would take place. Time must be "ripe," they would say. Before the dances begin, people wait for hours until all preparations in the *kivas* are completed. Time is important on these occasions, but it is not our time; it is rather the duration of certain ritual events relevant to Hopi life.[21]

Among the peoples of the Southwest, a sense of proper religious and social conduct prevails over any notion that time might be "usefully" employed.

Could such customs mean that people of other cultures experience the flow of time in a different way than we do? A number of distinguished anthropologists (among them such notable figures as Emile Durkheim, Claude Lévi-Strauss, and E. E. Evans-Pritchard) have proposed that the actual perception of time itself varies from culture to culture, depending on the particular system of time reckoning.

In this photo taken in 1916, a patch of sunlight shines down into the windowless sanctuary of a Hopi kiva through the hatchway in the roof. Similar lighting effects were sometimes used to time ceremonies.

However, if the basic human grasp of time were entirely relative, as this theory implies, it would surely play havoc with the normal logic and sequence of speech. We could not hope to penetrate the minds of ancient astronomers or of the calendar keepers of recent tribal societies. The thinking of other cultures would remain inaccessible to us, because it would be impossible to understand or express it through language. If a certain fundamental order were not shared by all languages, it would be impossible to communicate at all, no matter how able the translator. The people of other cultures do not literally "think differently" than we do, though to gain an understanding of their thoughts we may have to study the features of a social world very different to our own. It is surely common sense to suppose that a basic level of time perception exists that is common to all people, even though there are countless ways of interpreting and measuring it.

There is nothing mysterious, then, about the passing of time as experienced by the people of another culture. Their attitudes toward time are merely one aspect of a whole range of social customs that is unfamiliar to us. The intimate relationship between thinking about time and other social habits is illustrated well in the following account of the Pueblo of Zuni, New Mexico. When the noted southwestern anthropologist Elsie Clews Parsons visited Zuni in the early years of this century, only a few clocks had been bought by the inhabitants. Even these few, according to Parsons,

> . . . could fairly be described as ornamental. How little any other need for them is yet felt I realized one day in discussing the subject of Zuni meal-time with the proprietress of a clock. When the people ate only two meals a day, she said, breakfast and supper, they had no need of clocks. It was only when they learned from the Americans to eat a midday meal that they had to know the time. In this woman's mind the American meal and the American way of timing it were closely associated. An American habit was to be established in an American way.[22]

In an effort to understand other timekeeping systems besides "the American way," it is necessary to imagine what it would be like to live without clocks. We would soon lose our sense of time as a series of abstract quantities slipping into the past. Instead we would become conscious of the pattern of recurring events and how they shaped our lives. In fact, until recently, a preoccupation with repeating, cyclical events characterized the outlook of many Native Americans.

Since they had no clocks, telling the time during the day or night could be a haphazard affair. Although they devised simple sundials and also practiced watching the stars and constellations at night, many were content to describe the sun's position in the sky or the changing light conditions during the day in vague terms.

The Chippewa of Minnesota, when visited by an anthropologist during the 1930s, were found to be making some attempt to track the hours. They established a north–south line by setting up two sticks in the direction of the Pole Star during the night. The same sticks were then watched during the day, like a sundial, to establish noon and the times of two equal divisions of the morning and afternoon. Such devices would not have been useful to the Eskimo, many of whom lived in latitudes where the sun's hourly movement is less marked, and where the moon and stars are invisible for several months. In this setting, it was actually easier for coastal Eskimo to tell the time of day from the ebb and flow of the tides than from looking at the sun.

Map of the North American tribes mentioned or illustrated in the book.

A few Native Americans kept track of the months with a calendar stick. Above, portrait of a celebrated Winnebago medicine man, Tshizunhaukau, probably dating from the 1830s. His calendar stick is now in the Cranbrook Institute in Michigan, and, according to Alexander Marshack, it seems to record the passage of the months in relation to the solstices. Right, a Pima photographed in the 1920s with a simpler type of recording stick, on which he had carved symbols reminding him of important events that had occurred during the year.

An awareness of natural cycles affected even the simplest level of time reckoning. The day was filled with a wide range of human activity, some of it predictable and some not, but the night—or sleep itself—was the most inevitable event of all. Few native people regularly kept track of time by referring to "days" (the Comanche, who reckoned by "suns," were one exception). Instead, "nights" or "sleeps" were the natural terms. The Navaho were greatly confused by the Spanish introduction of the day and the week, particularly by the fact that trading on a Sunday was forbidden, and this was perhaps reflected in their adoption of the Spanish word for Sunday, *domingo*, to refer to the day in general.

Recurring events also shaped the long-term awareness of time. "Winter" or "snow" (or, farther south, "summer" or "heat") were obvious terms for the "year" among the many peoples whose calendars were so imprecise that they were not fixed and divided at midsummer or midwinter. Sometimes other seasonal names stood for the "year," such as the evocative word *cohonks*, used by the Algonquins of Virginia to refer to the migration of wild geese that opened the winter.

Because of our printed calendars, we are accustomed to the abstract notion of a fixed length for the year. The importance of such a time unit was not so obvious to Native Americans, many of whom planned their activities by waiting for seasonal events such as the first snows or the ripening of berries.

While a rigid concept of "the year" was rare among North American tribes, so, too, was a purely seasonal year—one entirely dependent on the unpredictable arrival of migrating caribou or on the falling of leaves. Few communities, whether they relied mainly on hunting or on agriculture, were content to live from one season to the next in a state of total uncertainty about time. Instead, to provide some degree of predictability in their lives, the Native Americans nearly always related the arrival of natural phenomena to the sun or moon, or both.

THE MOON CALENDARS OF NORTH AMERICA

The rhythm of the waxing and waning moon was the simplest of all timekeepers. By naming each month after some recurring event in nature, the observers provided themselves with an elementary but useful calendar. In a summary published in 1919, seasonal names for the months were listed for more than sixty North American tribes.

Despite the importance of the moon to so many different peoples, the care and accuracy with which the months were followed varied enormously. John Murdoch, a member of a polar expedition that ventured into arctic Alaska during the early 1880s, found that this was true even within a single community. The Eskimo who lived in large villages at Point Barrow evidently did not observe the moon very consistently:

> We obtained with some difficulty the names of nine moons or lunar months of the year, but were told that for the rest of the year "there was no moon, only the sun". Dr. Simpson, however, obtained names for all twelve moons. The names of these moons were given differently by different informants, and do not wholly agree with those given by Simpson. It is quite likely that they are not invariable, and may be going out of use. [23]

This account illustrates the problems connected with our knowledge of these simple lunar calendars. The original researchers "on the spot" a century ago found it hard to be sure whether the disorderly month lists reflected genuine confusion or carelessness in the minds of the moonwatchers, or merely the decay of some system which had once been more refined.

In the arctic latitudes, where the moon was often invisible during the endless daylight hours of the summer, there was a practical obstacle in the way of accurately tracking all the months. Farther south, however, there was often just as much uncertainty. Some tribes of the Cheyenne, for instance, named twelve moons, while others troubled only with six; indeed, even within a single Cheyenne tribe, there was no uniformity in the seasonal names adopted for the months. A "blank" period of the year, when no moons were named, was characteristic even of some peoples who made precise observations of the sun and

Eskimo woman at Point Barrow, Alaska.

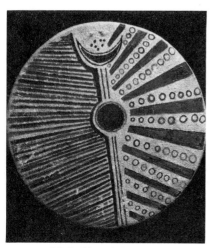

A model shield made by a Kiowa of the Great Plains.

followed a highly elaborate round of ceremonies, like the Bella Coolla and the Kwakiutl of the Northwest Coast.

Whether by neglect or deliberate custom, the Native American skywatchers who failed to count all the moons in the year were actually saving themselves a great deal of trouble. If every moon were accurately followed, a discrepancy would quickly arise. The source of the problem was the fact that an exact number of lunar months cannot be "fitted into" the solar year. If twelve months are chosen, they fall short of the actual length of the year by about eleven days. If, on the other hand, thirteen is the number selected, the year is too long by about nineteen days.

Imagine the case where twelve months, each named after seasonal events—the ripening of berries or the movement of caribou, for instance—were scrupulously followed over the course of a single year. By the time the next year came around, a

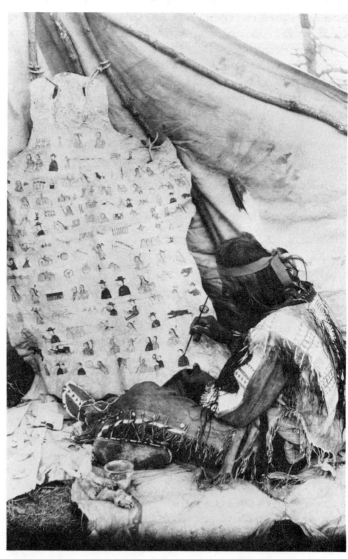

Sam Kills Two working on his Winter Count at the Rosebud Sioux reservation, sometime after 1910. The Winter Counts were pictorial histories, each symbol representing a notable event for that year. These events were sometimes sky phenomena, such as eclipses and meteor showers. Most Winter Counts date from the last century, and were perhaps an imitation of European habits of recording.

particular month would be well ahead of the actual seasonal occurrence it was named after. By contrast, with thirteen as the estimate, the caribou month would probably arrive well after the caribou themselves. The Dakota, who knew the months by such signs as laying geese, ripening strawberries, and raccoons stirring from winter hibernation, were often confused because their twelve moons never brought them back to their starting point.

Here was the classic difficulty which beset the calendar makers of all early civilizations. The cycle of the moon could not be matched exactly with that of the sun (and with phenomena related to the sun, such as the changing seasons, vegetation, and so on). For us, of course, it is easy to understand this simple discrepancy between the sun and moon because of our idea of a fixed length for the year which was passed down to us by the Egyptians and the Romans. However, for a people accustomed to watch for recurrent events rather than to think in abstract time units, the effect must have been perplexing. What efforts did the Native American skywatchers make to overcome their calendrical confusion?

At this point, it is perhaps necessary to repeat that the calendar keepers of many tribes simply neglected to follow the months during part of the year and thus avoided further complications. Even so, the question of which moon was which often became the subject of heart-searching, disagreement, and even open quarrels among tribal leaders.

This was the case among the Salteaux Indians, an Ojibwa-speaking tribe of hunters and fishers who lived in a permanent settlement on the Berens River near Lake Winnipeg. In the middle of the last century, a visitor to the Salteaux, J. G. Kohl, observed that "it is often comical to listen to the old men disputing as to what moon they are in."[24] The outcome of the disputes was that an occasional thirteenth month with no name was added to the regular sequence of twelve, which were called after such seasonal events as the arrival of eagles and the gathering of wild rice. The Salteaux made no attempt to time ceremonies so that they took place at a particular time during a particular month. Some individuals knew under which moon they were born, but others did not. This uncertainty even found its way into a Salteaux legend about the murder of an old man; the victim "is asked what moon it is and replies 'Midwinter Moon.' 'No, you're wrong. This is Eagle Moon. Look! There is an eagle now passing behind you.' When the old man turns to look, his throat is cut and the murderer remarks, 'Did you expect to see an eagle at this season?' "[25] This story suggests that disagreement about the calendar was a customary facet of Salteaux life. Provided such disputes did not seriously threaten tribal authority or disrupt the round of subsistence activities unduly, there was, indeed, little incentive for anyone to pursue the notion of a well-regulated calendar. An "agreement to disagree" was one way in which a lunar calendar could be kept in step with the sun, in an

Astronomical symbols from various Winter Counts of the Dakota Sioux.

Loud, brilliant meteorite, winter 1821–22.

Meteor showers of November 12, 1833.

Total solar eclipse, August 7, 1869.

erratic but workable fashion. Why should anyone have believed that astronomical events were any more reliable or predictable than the arrival of eagles or the dropping of leaves in the fall?

In an influential study published in 1978, two British academics, David Turton and Clive Ruggles, demonstrated that a constant state of disagreement about a simple lunar calendar may nevertheless result in an acceptable timekeeping system. Their study involved fieldwork carried out between 1968 and 1974 among the Mursi, a small, highly mobile group of cultivators and herdsmen in the Omo basin of southwestern Ethiopia. The seasonal activities associated with each of the thirteen months are well known to all the Mursi, but constant discussion and correction of the calendar take place. This is a matter of interest to ordinary herdsmen as well as to the informally trained diviners or fortune tellers, whose special skills include predicting the future by examining animal entrails. An extra month slips in or drops out of the sequence in a haphazard way as a result of the arguments. Because of this uncertainty it is impossible to pin down the date of a particular event more accurately than within about four weeks.

Nevertheless, the Mursi know the precise time of the all-important flooding of the Omo River, an event that signals the period to begin planting sorghum and maize along the riverbank as the waters recede. The Mursi recognize this time by keeping track of the successive setting positions of four stars in the area of

A wedding celebration of Mursi herdsmen in southwest Ethiopia.

the Southern Cross. When the four stars eventually set in the solar twilight and so become invisible, the time for the flood is at hand. In addition, observers also know the time that the sun enters its "house" on the horizon at midsummer or midwinter. None of these astronomical events is used as a means of settling the disputes about the state of the moon. Instead, the arrival of the solstices and the stars' changing positions are apparently regarded as no more predictable than any other seasonal happenings; it is all a matter of uncertainty, open to disagreement.

The study of the Mursi holds a vital lesson for anyone concerned with ancient astronomy, as Turton and Ruggles point out. Should an investigator discover traces of an alignment to the sun at a ruined monument, it does not automatically mean that the builders kept a highly organized calendar. The sun may have been watched for the sake of a special belief, or to signal a particular event, while at the same time an informal, often chaotic, sequence of lunar months was followed and debated. The need for an accurate, objective standard of time measurement, so vital to our lives today, played little part among many of the societies discussed in this book.

THE PAWNEE, STAR PEOPLE OF THE PLAINS

The records of North American tribes reveal that there were several different ways of resolving disputes about the calendar, if indeed it was considered necessary to do so. An awareness of the stars often played a part, since the pattern of the constellations shifts in tune with the seasons. This was a change that many North American chiefs and priests observed intently. They would sometimes use the first annual appearance of a star in the dawn or evening sky as a sign to start a particular month or an important ceremony.

Among the most skillful star watchers in North America were the members of the Skidi Band of the Pawnee tribe. The Skidi Pawnee originally lived in villages located along the Loup and Platte rivers in eastern central Nebraska, before the tribe was removed to Oklahoma toward the end of the last century. Their society was characterized by a complex hierarchy, and they pursued a varied round of subsistence activities ranging from sedentary maize growing to mobile buffalo hunting. They also acquired a deep knowledge of the stars.

The astronomical interests of the Skidi Pawnee were overlooked when the first ethnographic account of the tribe was compiled during the last century. J. B. Dunbar, the son of a Presbyterian missionary who had worked in the Pawnee villages in Nebraska during the 1830s and 1840s, commented that "their knowledge of astronomy was very limited. . . . Of the stars they discriminated few by name."[26] Or so Dunbar claimed. When ethnographers such as Alice Fletcher and James Murie studied the Skidi on their Oklahoma reservation around the turn of the century, they found, by contrast, that the Band's mythology, rituals, and social

divisions were all heavily influenced by beliefs about the stars. In fact, the Skidi had developed more elaborate star ceremonies than any other tribe on the Great Plains.

Today, over a century after their removal to Oklahoma, the full extent of Pawnee astronomical knowledge is still in the process of being recovered. Recent research by the Smithsonian astronomer Von Del Chamberlain, who has studied the field notes of early observers, has revealed a wealth of new information.

The Skidi Pawnee were convinced that the stars represented supernatural beings who maintained vital relationships with people on earth.

Like the members of Skidi society, the star gods were ranked in order of importance. First came the red Morning Star warrior (probably Mars), who mated with the female Evening Star (Venus) to produce the first humans. Next came the four gods who supported the heavens, located at the semicardinal points (northwest, northeast, southwest and southeast). The sun, moon, and gods of the cardinal directions were all of lesser importance, although a first male child was said to be the offspring of the sun and moon. The North Star was pictured as the chief watching over all the other stars and people on earth. There was also a constellation called "The Circle of Chiefs," which lent sanctity to the Skidi's elite Chief's Society on earth.

The ultimate source of power for all these stellar beings was the central force of the universe called *Tirawahat*, an invisible presence in the sky. It was Tirawahat who endowed the people with a bundle of sacred objects to be kept at the shrine of the Evening Star in the west. The other star gods followed this example by depositing holy bundles at their own shrines. By establishing this contact, the star gods transmitted "the power to put life into all things, to set the people in order, and to give them knowledge."[27] (So a Skidi informant told Alice Fletcher at the turn of the century.)

This highly ordered cosmology served as a kind of blueprint for the organization of Pawnee life on earth. The villages of the Skidi Pawnee were said to have been laid out in a pattern duplicating the positions of the most important star gods in the sky. Four important villages were grouped in a square at the center. Alice Fletcher's local informant explained the significance of the central villages as follows:

> The Skidi were organized by the stars; these powers above made them into families and villages, and taught them how to live and how to perform their ceremonies. The shrines of the four leading villages were given by the four leading stars, and represent those stars which guide and rule the people.[28]

An outlying village at the east contained the all-important shrine dedicated to the Morning Star, and another at the west contained that of the Evening Star. The other, lesser villages were grouped around the center according to the position of their patron stars in the sky.

The traditional dwelling of the Skidi at their permanent villages was the earth lodge, which was constructed so that its layout conformed to the heavenly order above. The circular, partly subterranean lodges, usually less than fifty feet in diameter, were built to face east, so that during part of the year the sun's morning rays would penetrate through the low tunnel entrance and strike an altar set up against the western wall. A central fireplace represented the power of the sun. Around it were positioned the four great posts that supported the wooden superstructure of the lodge. This overarching, domed wall and roof symbolized the sky. The four main posts were set at the semicardinal points and painted in the colors appropriate to the star gods of the four principal villages. The sacred medicine bundle, which would usually contain such items as ears of corn, animal skins, or smoking pipes, was hung up on the wall near the altar at the west. With the proper ritual prescription, the power of this bundle could be invoked each year to ensure the well-being of the people. Both the tunnel entrance and the smokehole situated

Opposite, the Pawnee leader Raruhca-kure:sa:ru, "His Chiefly Sun," photographed by Jackson in 1871.

Opposite below, Pawnee earth lodge in a village on the Loup Fork, Nebraska, photographed by William H. Jackson in 1871. View inside a Pawnee earth lodge, below, shows the low earthen altar, positioned against the western wall diametrically opposite the entrance to the lodge.

over the hearth were apparently sometimes used as observing devices to track the seasonal shift of the stars. By installing himself on the floor at various positions around the hearth, a priest could make sightings of the Pleiades and other constellations that were of significance to the Skidi Pawnee.

The records of their calendar compiled toward the end of the century make no reference to the solstices. Instead, the year began with the spring ceremony of reawakening, when the Skidi had returned to their settlements after the exhausting winter buffalo hunt on the Plains. The timing of the ceremony depended initially on the appearance in the southeast sky of two small stars known as the "swimming ducks," whose duty was to arouse all the other creatures from their long sleep. This period was also judged from the position of the Pleiades in the sky. In this way, the appearance of the stars must have been used to set the moon count straight, although a twelfth month of *kaata,* or darkness (occasionally paired with an extra thirteenth month), ensued before the year officially began. During this time, the priests waited anxiously for low thunder on the distant horizon, which was thought to be the voice of heaven, for only with this sign could the elaborate spring creation rituals begin.

Despite the precision with which the Skidi attempted to adjust the starting point for their sequence of months, the question of how many months made up the year still provoked disagreement. Gene Weltfish, who lived with Pawnee elders in Oklahoma during the 1930s, has reconstructed such a discussion, set in the firelit earth lodge of earlier times:

> It was in just such a meeting right after they returned from the winter hunt, after the warriors and chiefs had told their stories, that the old men would take up the question of the calendar and whether they needed to introduce the thirteenth intercalary month that year. . . . Some said, "Not now," others thought this was the time. They spoke of the position of the constellations, "first one snake, then two ducks and then the real rattlesnakes (Scorpio). That marks the beginning of the year—the spring. The months should coincide with those constellations."[29]

In the earliest ethnographic account of the Pawnee, published in 1882, J. B. Dunbar observed that the temperature of these debates occasionally rose:

> They sometimes became inextricably involved, and were obliged to have recourse to objects about them to rectify their computations. Councils have been known to be disturbed or even broken up, in consequence of irreconcilable differences of opinion as to the correctness of their calculations.[30]

In many cultures, the first annual appearance of a constellation in the dawn sky was a signal for the beginning of the new year. In this fanciful engraving, Samoans joyfully greet the arrival of the Pleiades in the midwinter sky.

The "objects" mentioned by Dunbar were apparently notched sticks, which he says recorded the passage of nights, months, and even years. The sticks were also carved with symbols: a star standing for the sun, a cross for the night, and a crescent for the moon or lunar month.

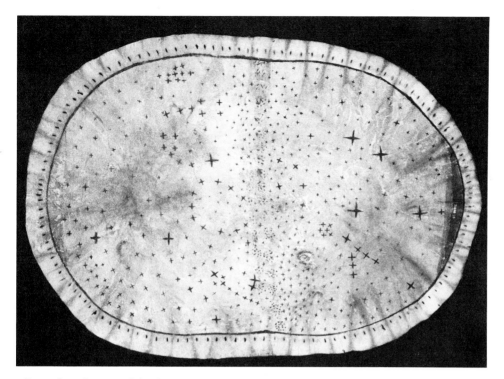

Recording devices of this kind were the obvious first step toward acquiring a more accurate framework of time reckoning, but many North American tribes pursued a successful existence without them. Indeed, exact standards of timekeeping were unnecessary even for a people like the Skidi, who were preoccupied with star lore to the point where it dominated their mythology, the organization of their settlements and social lives, and even the interior arrangement of their homes. They carefully observed the movement of the stars both at the horizon and overhead. Despite these astronomical skills and ritual concerns, they were apparently content with a comparatively crude lunar calendar, even though it provoked considerable disagreement.

The lesson of the records of the Pawnee and of many other North American tribes during the last century is clear. They developed an elaborate sky mythology and a detailed knowledge of celestial movements, yet felt no compulsion to fix a rigid, regulated calendar. Indeed, their thoughts were freed of the preoccupation with measuring linear, abstract time so essential to our own culture. Uncertainty, disorder, and a lack of concern with defining regular units of time all characterized the attitudes of Native North Americans to their calendars.

A remarkable Skidi Pawnee star chart, painted on a buckskin in at least three different pigments. The chart shows most of the important constellations identified by the Pawnee. This symbolic document, perhaps a century old, was kept in one of the sacred bundles that the Pawnee believed had been transmitted by the star gods to people on earth.

9. "Crystals in the Sky": The Astronomy of the Chumash

The achievements of North American skywatchers would not excite much interest if they had been limited to informal, chaotic methods of timekeeping. In certain regions, however, particularly in the Southwest and along the West Coast, the situation was different. Here, many diverse tribes were organized in more complex social patterns than existed elsewhere in North America. With the emergence of powerful hierarchies and secret cults came a more formal development of calendar skills and a deepening religious preoccupation with the night sky. Here, among the wealthy, competitive chiefdoms of the West Coast, the astronomer became a revered and often feared individual.

The question of how and why complex societies developed has intrigued anthropologists for many years. An answer to this question would obviously shed light on how astronomical knowledge changed from a loose and irregular basis, as we have seen in the disputes about the lunar calendars, to become more exact and systematic.

One of the most popular theories is that the crucial step forward was the invention of agriculture, which occurred about 7000 B.C. in the Near East and about 5000 B.C. in Mexico. Farming is supposed to have granted prehistoric people more leisure time to devote to their skywatching skills, and enabled their leaders to exercise greater power by controlling food resources that could be stored and traded easily. Some scholars of the ancient past suggest that the annual routine of agricultural sowing and harvesting led to the invention of the calendar; this encouraged a steadier, more orderly outlook on natural events in general. In this way, farming is held responsible for the growth of societies beyond a simple egalitarian level, and also, indirectly, for arousing man's spirit of scientific inquiry which is so influential today.

But recent studies of Native American astronomy show that many of these assumptions are mistaken. The records of the Californian tribes are particularly exciting, for they prove that nonagricultural peoples could develop far in the direction of social and intellectual complexity. In fact, it is clear that the capability of Californian hunter-gatherers to make sustained observations of the sky was in no way inferior to that of their counterparts among any other North American groups, including those who raised corn.

Of the three hundred or more indigenous groups, or "tribelets," of California, the Chumash were, until recently, one of the least well known and understood, even though they were the first to be encountered by Europeans. On October 10, 1542, the Spanish adventurer Juan Rodríguez Cabrillo sailed into the Santa Barbara

Channel, landed at Ventura, and subsequently explored both the coast and the outlying islands. The journal of his voyage records that the Indians were clad in skins, their long hair tied up in elaborate styles, and that they occupied permanent settlements ranging from hamlets to substantial towns. Their spacious semi-circular houses, solidly constructed from frames of wooden poles thatched with grass, won the admiration of the early Spaniards, together with the natives' intelligence and industry. The author of the Cabrillo journal reported that they were organized in two great provinces, one of which at that time was ruled by an old woman.

There may have been as many as 10,000 people living along the fringes of the Santa Barbara Channel, one of the most dense centers of population anywhere in the New World. By 1831, only 2,788 Chumash were left, or at least only that many were registered at the five Franciscan missions established within their territory. Many had fled to the interior to escape the tyranny of the mission system, which in a mere sixty years had decimated the Chumash population through a combination of harsh treatment and the spread of infectious diseases. Today about twenty Chumash families occupy a recently built housing development on the tribe's small reservation near the Santa Ynez Mission northwest of Santa Barbara. Several thousand Chumash of mixed descent live elsewhere in southern California.

With comparatively little documented history available, only a handful of scholars were attracted to the study of the Chumash before the late 1950s. Then it became widely known that vivid multicolored rock paintings decorated the surfaces of small caves and overhangs in the remote chaparral country of the interior. A local commercial artist, Campbell Grant, began to visit and document the painted rocks on behalf of the Santa Barbara Museum of Natural History. To do so, he had to hike through almost impenetrable undergrowth to the wind-eroded sandstone nooks and crannies where the paintings are found.

Most of the designs consist of simple geometrical patterns or strange animal figures bristling with sticklike arms and legs. A

A mid-nineteenth century depiction of Native Californians with elaborately dressed hair.

few painted panels are truly spectacular: flamboyant, intricately patterned bands, zigzags, crosses, wheels and sunbursts appear alongside fantastic snakes, insects, and other creatures. The motifs are executed in bright mineral pigments, mainly in shades of red, black, and white. Their precise age is unknown, although some decorated panels give the impression that they could be quite recent. For example, there is a suspicion, as yet unconfirmed by scientific tests, that one of the most brilliantly colored of all the designs incorporates pigments of European origin.

The effect of Chumash rock art, especially when presented by Campbell Grant in his own accurate copies of the original paintings, is both impressive and disturbingly dreamlike. In his landmark study, *The Rock Paintings of the Chumash*, Grant's meticulous, glowing illustrations demonstrate that the Chumash created the finest rock art in North America. Yet beyond the guess that these hidden caves were the haunts of shaman-artists bent on working magic—possibly inspired by hallucinogenic visions—Grant found little hope of understanding their meaning.

But a few years after the publication of Grant's book in 1965, Chumash scholars realized that a rich source of new information lay almost untouched in the vaults of the Smithsonian Institution

Below and opposite, magnificent multicolored Chumash rock paintings at Painted Cave, near San Marcos Pass in Santa Barbara, California.

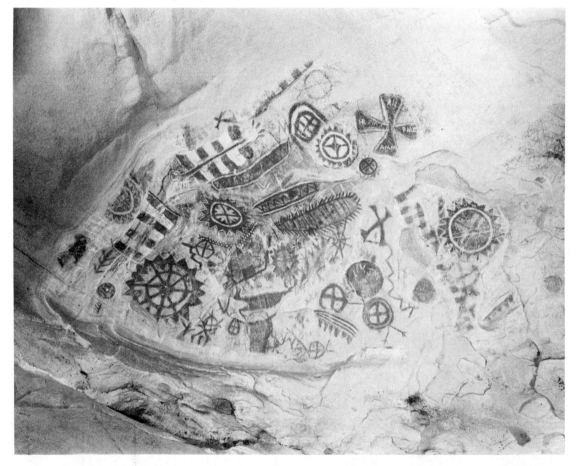

in Washington. The information consisted of an immense quantity of notes compiled by John Peabody Harrington, an anthropologist on the staff of the Smithsonian. From 1906 onward, Harrington devoted much of his life to recording the reminiscences of a number of elderly Chumash in the vicinity of Santa Barbara. Together, their recollections covered many aspects of traditional Chumash life, such as the myths, historical incidents, social arrangements, and technical skills which had belonged to the world of their forefathers.

Harrington accumulated a hoard of evidence with an unflagging, obsessive energy. A recluse, he spent much of his later life in a Santa Barbara hotel, working late into the night on the task of ordering his notes. After his death in 1961, some eight hundred thousand pages, stuffed into more than one thousand boxes, eventually found their way to the Smithsonian archives. (There were about sixty boxes on the Chumash alone).

Today the full magnitude of the Harrington legacy is still not properly known, since the effort of researching and publishing his notes is almost as formidable as the energy he consumed in collecting them. One dedicated Harrington scholar, Travis Hudson of the Santa Barbara Museum of Natural History, has already

John Peabody Harrington's devotion to recovering the Chumash way of life is evident in these poses for the Smithsonian, displaying a Chumash throwing stick, a tule (or bulrush) mat, and a bow and arrow.

compiled several illuminating studies of Chumash life. The most impressive of these studies is a book coauthored with an astronomer, Ernest Underhay, evocatively entitled *Crystals in the Sky*. Even for experts in the field of ancient astronomy, the appearance of this work in 1978 came as a revelation. The detailed accounts of cosmology and sky myths extracted from Harrington's notes, combined with field studies of rock paintings and possible observatory sites, resulted in a remarkably complete picture of Chumash astronomy. *Crystals in the Sky* showed that native Californians possessed a more complex outlook on the universe and far greater powers of observation than had ever been suspected.

A POWER STRUGGLE IN THE COSMOS

Their fundamental belief was that a balance of supernatural power existed in the world. This was an all-embracing power to which humans were subjected, yet which they could also influence for their own good or ill. Human affairs did not occupy the center of the stage, however. Since earliest times, a hierarchy of much more powerful beings had been responsible for changes on earth and for the destiny of people. The underworld was much feared, for here lurked malevolent spirits. The middle world—the flat, circular earth, surrounded by a void or by water—was inhabited by supernatural entities as well as by people. The most important forces of all lived in the upper sky world, where they were locked in an intensely competitive struggle for control of the universe. The sky beings were actually people who had ascended from earth long ago to escape death in the primeval flood, and so possessed some human qualities; their unpredictability and fickle character made their power fearsome. But humans were not helpless, for they could enlist the aid of supernatural forces in various ways. Spirit helpers might appear in dreams, the services of a shaman could be engaged, and there were also public ceremonies that affected the cosmic balance of power. By these means, the threatening intentions of any particular celestial being would not get out of hand and destroy the world. In short, the Chumash outlook emphasized the importance of supernatural forces, and viewed man's role in the universe as an agent who could affect those forces for either good or ill.

Daily rituals and observations were essential for the Chumash priests to interpret the actions of the sky beings. These beings seem to have been ranked in a hierarchy, as was Chumash society itself. First came the sun, believed to be an aged widower living in a quartz crystal house, who carried a blazing torch on his daily journey across the sky. The moon was a female who controlled human health and the menstrual cycle of women, and also seems to have been identified with *datura*, the powerful hallucinogen drunk by shamans to perform cures and foretell the future. The twin appearances of Venus seem to have been treated separately by the Chumash. The Morning Star was benevolent, perhaps associated with rain; in contrast, the Evening Star was the feared chief of the underworld, probably the giant golden eagle who removed and ate the bones of people from the middle world. Another celestial bird, the California Condor, possessed magic clothing that allowed him to locate missing objects or persons. This deity, who also carried two sticks that allowed him to jump quickly across the sky, seems to have been the planet Mars. Jupiter and Saturn may also have been identified, although Harrington's notes do not clearly confirm this possibility.

In addition, many stars and constellations were regarded as important beings—notably the North Star (probably the benevolent Sky Coyote) and the Big Dipper (seven boys who were transformed into geese). It is clear that an enormous number of stars must have been named by the Chumash and other Californian peoples, but most of their lore was never recorded. The Luiseño of southern California, for instance, are known to have identified extremely faint stars of sixth magnitude as parts of their mythical sky beings or constellations.

Every important event in the calendar presented a crisis during which the struggles of these supernaturals were observed and action was taken to avert a potential catastrophe. Each month the moon had to be restored to life, which the Chumash and other groups did by shouting encouragement and extending their arms in prayer.

The winter solstice was the most critical moment of all because of the possibility that the sun might choose not to return. It also marked the annual finish of a nightly ball game played by two teams of sky people, one led by the sun and the other by Sky Coyote (the North Star), with the moon acting as scorekeeper. The game was an opportunity for the most powerful celestial beings to assert their influence and so upset the balance of nature; the outcome was literally a matter of life and death. It was

Sketch of the sunstick from Bowers Cave, Los Angeles, "discovered" in collections at the Peabody Museum by the author and Travis Hudson. The stone, with its traces of red-painted "rays," is about four inches in diameter.

essential for Sky Coyote, the benefactor of people on earth, to beat the sun's forces.

Human participation in the cosmic struggle reached a climax in the ceremonies of midwinter, which involved the entire community. The public observances lasted for several days, beginning with a gathering of those who had incurred debts over the course of the year. A share of the shell money that changed hands at this event enriched the local chief, the *wot*, who in turn subsidized the ceremonies and provided help to the poor. The next day the high chief known as the *paha* put on the robes that proclaimed him to be the "Image of the Sun." His twelve priest assistants, who belonged to an elite cult called the *'antap*, impersonated the sun's rays. The officials then stood a sunstick upright in the ground. This was a small ritual object which Harrington's notes describe as a wooden shaft crowned with a painted stone disc; the shaft apparently represented the axis of the world, while the stone was the sun. At both solstices the *paha* positioned the sunstick in such a way that only the shadow of the disc fell on the ground. He then struck it with a magic stone and uttered an incantation intended to "pull" the sun back toward the earth.

Two sunsticks are currently on display at the Peabody Museum of Harvard University; a third was recently examined in the museum's stored collections by Travis Hudson and myself. The objects seem to confirm the account of Harrington's informants, for each disc is fixed at a skewed angle on the shaft, similar to the angle of the sun at midsummer noon if the shaft were positioned upright. The painted lines on the top surface of the best-preserved Peabody sunstick are also interesting; if, indeed, it was oriented as described in the native account, then the red-painted "rays" point roughly in the direction of south and the solstice positions. Incidentally, one of the Peabody specimens has about 150 notches engraved on the shaft, possibly representing a tally of the days leading up to the solstice.

The midwinter ceremonies continued for the following two days and nights. Among the activities were dances dramatizing the soul's journey along the Milky Way to reach the land of the dead. The people also erected great feathered poles, which were eventually moved to sun shrines located on the hilltops.

These shrines may also have been used for watching the sun, since there is no doubt that solstice observations *were* conducted throughout California. In the Chumash documents we read of one old man who, at midwinter,

> . . . would watch the rising sun, seated in front of his door, on the ground. Three peaks can be seen to the south of east from his position. The sun would pass the middle peak on the way south, pass the valley, remain two days, and on the third day would come up again over the middle peak on its way north. He would notify the other Indians of the New Year. He followed this practice for many years.[31]

This particular example of sunwatching would obviously not

leave any traces behind for archaeologists to identify, and so the task of recognizing the remains of observing sites is not straightforward.

Nevertheless, with the new wave of interest in Californian astronomy stirred by the publication of *Crystals in the Sky*, special efforts are currently in progress to identify solstice observatories throughout the state. At least fifteen promising sites have been discovered in central and southern California. This research is at a preliminary stage, since the alignments to the solstice have all been found as a result of on-the-spot observations, and have not been confirmed by precise measurements with a surveying instrument. Furthermore, some of the observatories seem unconvincing; in one case, the only indicator of a sight line to a distant peak is a painted rock design that *may* depict the sun above a mountaintop. This kind of evidence would not impress the careful investigator familiar with the problems of megalithic astronomy.

On the other hand, several of the Californian sites may well have incorporated symbolic, indirect alignments to the sun. In Baja California, I was guided to a small painted cave by Ken Hedges, an anthropologist at the San Diego Museum of Man. Hedges told me how, on midwinter day in 1975, he had witnessed a striking play of early morning light among the painted figures. It reached a climax when a thin streak crossed the eyes of a red-horned human, perhaps a shaman wearing a mask.

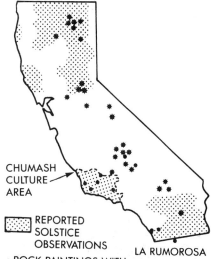

CHUMASH
CULTURE
AREA

LA RUMOROSA

⬚ REPORTED SOLSTICE OBSERVATIONS

✱ ROCK PAINTINGS WITH SOLAR MOTIFS

• POSSIBLE SOLSTICE OBSERVATORIES

Map of Californian sites, showing areas where solstice observations were reported by anthropologists during the last century, as well as recently identified sites where such observations may have taken place.

The figure of a shaman painted in red on the wall of a rock shelter at La Rumorosa, Baja California. At midwinter a pointed light form penetrates the shelter and crosses the face of the shaman.

Midwinter sunrise viewed from a rock alignment that may have been an observing site at Cowles Mountain in San Diego County, investigated by Ken Hedges of the San Diego Museum of Man.

While the evidence of possible solar observatories requires further study, there is no doubt that the Chumash possessed other aids for keeping track of their calendar. For example, the start of the twelve months of the Chumash year was timed according to the solstices and to the rising of important stars on the eastern horizon. The astronomers used star maps with patterns of shell beads inlaid on a background of tar so that the appearance of the constellations could be recognized. (Incidentally, the connection that each month had with a special star being in the upper world also provided the Chumash with a form of astrology; the list of qualities associated with people born in each month reads like a modern newspaper horoscope).

Besides the star maps, the Chumash and other native Californians used such devices as tally cords, notched sticks, or threaded beads to keep track of astronomical cycles. The Pomo of northern California apparently kept records stretching back over 500 years. The head of their secret society was reported to have used sticks to mark the passing of the months; a bundle of thirteen sticks represented the year. These, in turn, were grouped in three larger bundles of 8, 64, and 512 years, respectively. If the Californians were able to maintain records systematically for so long, it is conceivable that they gained an understanding of the appearances and disappearances of Venus, or even of the timing of eclipses. Such a possibility is totally at odds with the conventional picture of the limited intellects of hunter-gatherers.

The rock art of the Chumash is a vivid reminder of how we have underestimated the creative abilities of such people. In the wake of *Crystals in the Sky* and other recent studies documenting Californian astronomical skills and cosmologies, researchers have naturally turned to rock art with a fresh eye. Symbols that look like suns, moons, stars, eclipses, or comets are now identified more confidently in the knowledge that Chumash astronomer-priests were intensely preoccupied with understanding celestial phenomena. Some of the peculiar animals depicted in the

A painted sun-like symbol at Edward's Cave, Los Padres National Forest, southern California.

painted scenes seem likely to represent specific powerful beings of the upper world such as Golden Eagle and Sky Coyote.

Harrington's obsessively thorough field notes throw light on the belief of Chumash shamans that they could communicate with supernaturals through the use of *datura* and other ritual aids, such as pipes, wands, and feathered poles. Indeed, it seems probable that some of the painted designs were executed as "magical" acts, intended to propitiate or control the beings depicted on the rocks at moments of crisis. In *Crystals in the Sky,* Hudson and Underhay conclude that the rock paintings were probably carried out by the astonomer priests

> . . . as part of a vision quest; it is likely that some of the more elaborate depictions contain astronomical motifs or subjects, particularly in view of the importance attached to the heavens by cult priests and shamans. Some paintings were created because the very process of depicting certain symbols activated supernatural power. [32]

Guided by the noted Chumash rock art photographer, Bill Hyder, I visited several decorated rock overhangs high up in the hills overlooking Santa Barbara. One panel consisted of a red sunburst with a tail attached to it, alongside the outline of a strange, phallic, lizardlike creature. I was struck by the resemblance of the sunburst design to a comet, a phenomenon that Harrington's informants said was greatly feared as a sign of imbalance in the upper world. I could easily visualize a Chumash

Chumash "comet" painting photographed by the author in the hills above Santa Barbara.

shaman-artist gazing at the rock in an intoxicated trance, hoping to ward off the evil omen by painting an image of the comet alongside that of his animal spirit helper. The ritual worked, for the comet passed and order was restored to the universe.

Meanwhile, back in the present day, a new crisis affects the continued existence of the painting. Vacant lots were being bulldozed for the erection of ranch-style homes within a stone's throw of the painting. There is no protection of any kind for Californian rock art, since virtually all of it is situated on private land. Vandalism and the indifference of landowners take an almost daily toll of the paintings, so they are disappearing fast. It seems ironic and disgraceful that we should be in danger of losing perhaps the most impressive traces of New World hunter-gatherers so soon after beginning to understand their full significance.

THE CULTS OF CALIFORNIA

Above and opposite, *sun-like designs from various Chumash rock paintings.*

The astronomers of the Chumash were not on the fringes of society, like wandering gypsy fortune-tellers today; on the contrary, they were vital to the functioning of a complex, competitive way of life.

The influence of the Chumash astronomer, or *'alchuklash*, extended into practically every aspect of sacred and secular affairs. His control over the calendar meant that he was involved with the naming of children, so determining their astrological character and destiny. He was also called upon when young men passed into manhood, almost always through the undertaking of a vision quest. To assist in the quest, the *'alchuklash* administered *datura* and supervised the search for a dream helper to be the young man's guide in later life. Other responsibilities typical of the parts played by shamans among many Native American peoples included the curing of sickness and the control of the local weather. There were rainmaking rites, or storms to be averted, sometimes by the execution of a "magic" rock painting.

But the interests of the *'alchuklash* extended far beyond these services to a single community. They were among the most respected members of a powerful, far-ranging cult organization known as the *'antap*. This was an elite religious society into which privileged children were baptized at an early age. Only the wealthiest class could afford the high expenses of the *'antap* office. They were further set apart from the Chumash commoner by the cultivation of their own refined, esoteric language.

Since a town chief and all the members of his family were required to be cult members, and there were usually twelve or more *'antap* within most communities, the priests exerted a strong influence. As interpreters of the outcome of the celestial ball game between the sun and Sky Coyote, the *'alchuklash* must have been regarded with particular awe, for the balance of the universe was at stake. Their astronomical lore served to reinforce the framework of society; the leader of each *'antap* group, the *paha*, was himself the "Image of the Sun."

California has always offered exceptionally abundant natural resources, but they are not distributed evenly across the landscape. Since there were once as many as three hundred thousand native Californians living in the state (complicated by their division into at least seventy-five different languages with a further three hundred dialects), there had to be an efficient mechanism of exchange linking one group to another.

In southern California, this was provided by ritual organizations such as the *'antap* cult. Their major responsibility was staging the key religious dances and ceremonies, which required consultations not only between separate villages and towns, but also at a

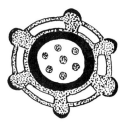

The formidable influence of the Chumash shaman is conveyed in this photograph dating to 1878.

provincial level. Their activities affected economic matters, since one of their duties was to collect foods that would be stored at the tribal capital for allocation to the needy by the chief, and to report on shortages that might occur in any local region. Religious rites were held regularly in many different locations and were always accompanied by food exchanges.

The most important of all was the Mourning Ceremony, held every two or three years to help the souls of the recently deceased on their way to the underworld. This event sometimes drew people by the thousands. We know that the Yokuts of southern California timed their Mourning Ceremony to coincide with the disappearance of the Evening Star, and it is likely that this was not the only case in which observations of the astronomer-priests served to coordinate ritual and economic activities.

Chumash astronomical lore, as it was interpreted by the powerful priest-shamans of the 'antap, penetrated every aspect of society and affected the life of every individual from birth. The belief in supernatural sky beings reinforced the Chumash social

Many Native Californians, like the Hupa photographed here at their White Deerskin Dance in the 1890s, belonged to rigid societies with a strong drive for wealth, property, and prestige. To become a wealthy Hupa, one had to acquire status goods like scarlet woodpecker scalps, albino deerskins, or the red obsidian knives carried by the chiefs in the foreground.

order, for the insecure balance of cosmic forces was thought to depend partly on the councils of the *'antap*. The ordinary Chumash commoner found himself in a highly restricted—indeed, almost feudal—situation; his dependence on the *'antap* officials must have been strengthened by the belief that they were responsible for keeping fearful celestial beings in check.

Astronomical conceptions formed the backbone of these rigid Californian societies. But was a social setting of this kind *always* essential for the growth of complex astronomical ideas and observations? Did every skywatching tradition develop because of a specialist class of priests who cultivated an exclusive knowledge of the heavens and commanded vast social influence? In a setting very different from California, was it possible for sophisticated astronomical ideas to emerge without the all-powerful priest figure?

The deserts of Arizona and New Mexico might seem much less favorable than the California coast, yet here, too, some of the best documented and most advanced North American systems of skywatching flourished. Furthermore, these traditional practices are alive today among such tribes as the Hopi and the Zuni.

A century or more ago, the societies of the Southwest were as complex and varied as those in California, yet in general their solutions to the problems of existence took a quite different path. Instead of an oppressive hierarchy, most communities in the Southwest had a fairly egalitarian structure, even if it was complicated by a mass of interwoven religious cults and family groupings. And although the astronomer-priest was a person who commanded attention and influence within the community, his power was limited. There was no need of an elite body to uphold his views, for at one time or another almost everyone participated in the most sacred dances and ceremonies. The status of the sun priest was a matter of respect and convention, rather than a distinction that could be measured in terms of wealth or upbringing. His knowledge of the heavens was not a carefully guarded secret, and was, in fact, open to criticism from the ordinary farmer. It was also accessible to the persistent questioning of anthropologists. This process began just over a century ago with one of the most unusual characters ever to study the American tribes.

10. The Sun Priests of the Southwest

One afternoon in the 1880s, a journalist was relaxing in the officers' club room at Fort Wingate, an isolated army outpost in western New Mexico, when his eye was caught by

> . . . a striking figure walking across the parade-ground: a slender young man in a picturesque costume; a high-crowned and broad-brimmed felt hat above long blonde hair and prominent features: face, figure and general aspect looked as if he might have stepped out of the frame of a cavalier's portrait of King Charles. The costume, too, seemed at first glance to belong to the age of chivalry though the materials were evidently of the frontier. There were knee-breeches, stockings, belt, etc., all of a fashion that would not have an unfamiliar look if given out as a European costume of two or three centuries ago.[33]

In this flamboyant outfit, Frank Hamilton Cushing, pioneer anthropologist, commanded the attention not only of the occupants of the frontier fort but also of the native villagers forty-five miles to the south at Zuni, where he had been sent in 1879 by the newly formed Bureau of Ethnology in Washington.

It was a novel idea for Cushing, then just twenty-two, to "go native" in pursuit of the secrets of Zuni society. Assigned for only two months, Cushing stayed on in the pueblo for five years, and his return was a reluctant one, forced on him by the unpopularity in Washington of his defense of Zuni land rights. By adopting items of Zuni dress, learning the Zuni language, eating a Zuni diet of mutton and suet, sleeping and shivering on a hard bed in the depths of the Zuni winter, Cushing hoped to win the respect of his hosts and to probe the innermost workings of their complex ceremonial activities. The results of his exploits appeared in a number of scientific journals, but it was Cushing's brilliantly woven popular narrative, *My Adventures in Zuni*, published in

THE PUEBLO SUNWATCHERS

Opposite, *Frank Hamilton Cushing (1857–1900) wearing the ceremonial dress he adopted during his stay at the Pueblo of Zuni, New Mexico.*

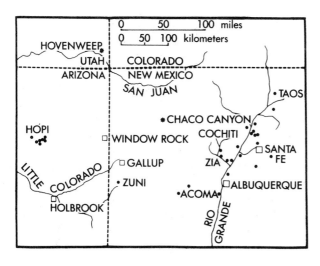

Map of pueblos (black dots) in the Four Corners region, also showing the location of the archaeological sites of Chaco Canyon and Hovenweep described in Chapter 11.

1882–1883, that brought his name and that of Zuni squarely before the attention of the American public.

Not all Cushing's colleagues were so tactful in their approaches to the peoples of the Southwest. A husband-and-wife team, Colonel James and Matilda Coxe Stevenson, with whom Cushing was originally dispatched to Zuni in 1879, provoked hostility with their aggressive efforts to obtain information. Indeed, at one of the villages of the Hopi tribe, "Tilly" Stevenson indignantly resorted to her umbrella in an effort to enter one of their sacred ceremonial *kivas*. This won her a place in the *Illustrated Police News* of 1886.

Earlier, at Zuni, the Stevensons were intent on collecting objects by the wagonload: household articles by the hundreds, along with ritual paraphernalia, dance costumes, hunting weapons, and harvest implements. All these objects were packed into boxes and hauled off under army escort through uncertain Navaho country to the nearest rail terminus, bound for Washington. A decade later, following the Colonel's death, the formidable Tilly returned to Zuni, her collector's lust unabated. This time it was information she sought—on a huge range of topics connected with Zuni life. The resulting six hundred-page volume *The Zuni Indians* remains the classic source book on the tribe.

Colonel and Mrs. Stevenson confront the Hopi in an engraving from the Illustrated Police News in 1886. The caption ran: "an assassin red-devil cowed by a white heroine."

Here was the heroic age of anthropology, when persistence, nerve, and physical courage were necessary to secure glimpses of a society still largely unaffected by modern American culture. Yet the insights won from the early study of the Southwest tribes carried a cost in local resentment that is felt to the present day. "The effect was not all good," says the guide recently published by the Pueblo of Zuni.

> The Zuni leaders felt that the anthropologists played up some things which were small and played down things which were important. The people believed that the very foundation of their culture was threatened by their activities. Many, like Cushing and Bourke, meant well, but still their main interest was the benefit of the Anglo culture. [34]

Today many Southwest peoples wish to protect their traditions from further inquisitive probing. This secrecy, together with the steady erosion of old beliefs under the impact of modern American culture, means that we must turn to the early anthropologists' accounts for an understanding of the sky lore of the pueblos.

These sources offer a rich insight into how a non-European people regarded natural events. We can trace the way in which astronomical practices were bound up with elaborate cycles of religious ceremonies affecting the lives of everyone in the community. Finally, we can pose the most vital question: what was the purpose of these traditions—the reason behind the entire complex system of observations and rituals? Was it all just a matter of irrational superstition, or did the pueblo dwellers' beliefs help them in some practical way to survive and prosper in their harsh desert surroundings?

Of more than twenty surviving pueblos in the American Southwest, the records of Zuni are particularly significant. This would be hard to guess from appearances, for today Zuni is an unspectacular place, a drab collection of low sandstone houses occupying a dusty rise beside the parched Zuni riverbed. However, its location among the remote, arid mesas of western New Mexico ensured that it was relatively unaffected by Spanish culture. While missionaries were busy reestablishing their grip on the eastern settlements of the Rio Grande, such as Taos, during the eighteenth and nineteenth centuries, the pueblos farther to the west were virtually left to their own devices until the coming of the Anglo-Americans in the mid-nineteenth century. As a result, through the eyes of Cushing and his contemporaries, we may glimpse in a relatively pure form the workings of a traditional desert culture.

To understand something of the part played by astronomy among the Zunis, we must first begin to appreciate their unusual social order. As in all the pueblos, authority in Zuni was never concentrated exclusively in the hands of a few powerful individuals; instead it was channeled through an extraordinary variety of religious organizations, constraints, and duties which molded the

"THE WAYS OF OUR FATHERS"

View over the rooftops of Zuni toward the sacred Thunder Mountain. Photograph by Francis K. Hillers in 1879, the year of Cushing's arrival in Zuni.

conduct of the individual from adolescence. The Zuni grew up with a deep sense of social and sacred discipline extending beyond the family. The interweaving of ceremonial ties bound all people together in a network that cut across individual family allegiances.

The traditional life of Zuni was organized through a bewildering variety of sacred organizations and ritual groups. The religious government of the village was in the hands of twelve *Ashiwanni*, or Rain Priests, four of whom were invested with special power as representatives of the four sacred directions of space. The most influential figure in guiding their decisions was the *Pekwin*, or Sun Priest, whose responsibilities in performing observations of the sun made him a particularly revered figure in the community.

Besides these officials, many other organizations flourished, such as the twelve curing societies and the famous *Katchina* cult. Today we usually think of *katchinas* as the wooden masked dolls that command high prices from art dealers. Originally, the dolls were made to instruct pueblo children about the many different *Katchina* spirits, who were thought of as ancestral beings and were identified with rain clouds. The members of the Zuni *Katchina* cult were subdivided into six groups, each with its underground *kiva*, where the holiest observances took place. Spectacular masked dances were also performed in public, and indeed many of these dances may still be watched by visitors to the pueblos today. To be initiated into any of these cult groups meant the mastering of a body of special rites and lore. Each cult contributed to an intricate web of sacred relationships.

In this society there was little room for personal nonconformity, since the success of a ceremony depended on the perfect ordering of gesture and song. Failure to live up to such responsibilities endangered the entire community, since periods of drought or destructive winds were the direct consequence of mistakes in ritual—an offering misplaced, or a prayer wrongly spoken. As in many societies where personal freedom is tightly restricted, hysterical outbreaks of witch-hunting accompanied natural disasters. Sorcery was punishable by torture and even death. As if to compensate for this regimented social conduct, the Zunis had an extraordinary variety of public ceremonies and societies, including suborders of clowns whose occasionally obscene play-acting was perhaps a safety valve for tensions within the community.

No Zuni rite, however, was more splendid or complicated than the annual *Shalako*, staged close to the time of the winter solstice. This elaborate and expensive night-long ceremony still flourishes today, drawing thousands of visitors to the village. The climax of the event is the blessing of newly built houses by the *Katchina* cult priest and the clowns, and by six "messengers from the gods," the great *Shalako* themselves, disguised under fantastic twelve-foot-high bird-headed effigies. Cushing vividly described these figures as "long-haired, bearded, great-eyed, and long-snouted, so managed by means of strings and sticks by a person concealed under

"The Shalako People," a painting by contemporary Pueblo artist Fred Kabotie.

their ample, embroidered skirts that they seem alive, and strike terror to the uninitiated."[35] One of the tasks undertaken by the *Pekwin*, or Sun Priest, was to ensure the ritually correct timing of the *Shalako* ceremony. It was supposed to coincide not only with the exact day of the winter solstice, but also with the night of the full moon.

Not surprisingly, the *Pekwin* usually had trouble in reaching a compromise between these demands, and his performance was a matter of critical concern for all. In 1896, for example, the newly installed *Pekwin* found himself in a timing dilemma that was resolved only after long discussions with the Rain Priests. His immediate predecessor had fared worse, for this man's mistake had been held responsible for a crop failure, and even witchcraft had been suspected. For his "bad heart," this earlier *Pekwin* was summarily dismissed by the Fertility Priestess. A similar sequence of events took place in 1915, when the outgoing *Pekwin* not only ignored the more or less permanent continence expected of his office, but tried to go by the American calendar and so mixed up his moons. Finally, in 1952, the Sun Priest evaded his responsibilities altogether by deserting the pueblo and moving to the city of Gallup, New Mexico.

The Zuni *Pekwin* had to anticipate the timing of the ceremony well in advance. For a period of eight days, he would prepare ritual prayer sticks of eagle feathers and undertake pilgrimages to the sacred Thunder Mountain overlooking Zuni; during this period he was believed to be in constant communication with his Sun Father. On the ninth morning, he would announce the approach of *itiwanna*: the solstice, or the sacred "middle" of the year, the focus of the winter observances. A further ten days were due to elapse before this event, during which all the Zuni people prepared offerings for their greatest ceremonial.

The tricky feat of timing was performed with the aid of observations of the sun as it rose each morning above the southern end of Thunder Mountain. The Sun Priest kept his watch from a petrified stump at the eastern end of the village. He would sprinkle the stump with holy cornmeal, and utter prayers as the sun rose each morning. The day on which he saw it reach a certain point on the mountainside was the time for his special duties to begin. The dawn observations continued on the final days leading up to the solstice, but thereafter no accurate watch was kept until the advent of midsummer.

Another account states that the midwinter sunrise was also viewed through a notch in the wall of a structure known as the "Sun Tower." When the sun lined up with this notch and with a pillar erected in the Zuni gardens, the signal was given for the solstice rituals to begin.

The ritual anxiety attending the sun's hesitation in his northern "house" at midsummer was much less than at midwinter, but the duties and observations of the Sun Priest were similar. Sunset observations were carried out from a little shrine situated on a rise

The Pekwin, *or Sun Priest, of Zuni, photographed by Mrs. Stevenson in 1896.*

near the ruined village of Matsakia, some three miles northeast of Zuni. Though it was a sacrosanct place, this primitive observatory was visited and photographed by Mrs. Stevenson. She described it as a low semicircular stone wall about a meter high and a meter across, open to the east. At the back of the wall stood a sandstone slab carved with a sun symbol. From this shrine, according to Tilly Stevenson, the *Pekwin* observed the gradual progress of the sun each evening as it neared its northern "house." This was a spot located on the mesa known as Great Mountain, to the northwest of Zuni. Sun Father was thought to rest there five times on successive days around the time of the summer solstice.

Dawn observations were also apparently made from the observatory at Matsakia or from a similar shrine nearby. Cushing describes them in a striking passage from one of his popular articles:

> Each morning, too, just at dawn, the Sun Priest, followed by the Master Priest of the Bow, went along the eastern trail to the ruined city of Ma-tsa-ki, by the river-side, where, awaited at a distance by his companion, he slowly approached a square open tower and seated himself just inside upon a wide, ancient stone chair, and before a pillar sculptured with the face of the sun, the sacred hand, the morning star, and the new moon. There he awaited with prayer and sacred song the rising of the sun. Not many such pilgrimages are made ere the "Suns look at each other,"

The sun shrine at Matsakia described in Cushing's account, here seen in a previously unpublished photo taken around 1920 by the archaelogist R. H. Lowie.

and the shadows of the solar monolith, the monument of Thunder Mountain, and the pillar of the gardens of Zuni "lie along the same trail." Then the priest blesses, thanks and exhorts his father, while the warrior guardian responds as he cuts the last notch in his pine-wood calendar, and both hasten to call from the house-tops the glad tidings of the return of spring.[36]

As at midwinter, the Sun Priest was traditionally bound to undertake eight days of fasting and retreat before giving public notice of the coming solstice. Cushing remembered this announcement well, for he was haunted by "the voice, low, mournful, yet strangely penetrating and tuneful, of the Sun Priest . . . heard from the house-tops."[37] The oration was believed to consist of words directly communicated to the *Pekwin* by Sun Father himself.

Cushing's testimony tells us that the delicate calculations of the *Pekwin* could be double-checked by ordinary people,

> . . . for many are the houses in Zuni with scores on their walls or ancient plates imbedded therein, while opposite a convenient window or small port-hole lets in the light of the rising sun, which shines but two mornings in three hundred and sixty-five on the same place. Wonderfully reliable and ingenious are these rude systems of orientation, by which the religion, the labors and even the pastimes of the Zunis are regulated.[38]

Such alignments may have been incorporated in other pueblos besides Zuni. At Cochiti Pueblo on the Rio Grande, an architectural feature reminiscent of the description of the Zuni Sun Tower was constructed during the present century. Through a slot high on the wall of his office, the religious head or *Cacique* of the pueblo could keep track of sunrises along the eastern horizon. Prior to the influence of Spanish calendars and customs, it seems probable that accurate sunwatching practices were as fundamental a part of life in the eastern pueblos as they were at the Zuni and Hopi villages in recent times.

Other devices served to jog the memories of religious officials and to keep the community in step with the sun. Knotted tally cords were kept by both Zuni and Hopi sunwatchers, and may well have been an ancient feature of Pueblo culture. Cushing believed that such mnemonic aids were once common at Zuni; in their language he found words for "song-strings," "tribute-strands," and "war-cords." Furthermore, he believed that the use of special fibers, colors, and styles of knot each carried a particular significance. Notched calendar sticks similar to the one carried by the Master Priest of the Bow to the sun shrine at Matsakia a century ago also assisted with time reckoning, although it may be that they had a fairly recent origin.

The existence of such recording aids, together with the aligned building features and the anxious watching of the sun along the horizon, all suggest that the Pueblo calendar keepers were concerned with precision and accuracy. From the dates mentioned by early anthropologists, we know that the timing of the major Hopi ceremonies was, in fact, repeated with remarkable consistency from one year to the next.

But the motivation of the sun priests was unlike our own contemporary urge for orderly measurement. They had no interest in fixing a neatly divided calendar with regularly spaced intervals, for their goal was not scientific understanding and prediction. Instead, they were preoccupied with following a traditional pattern for the organization of ceremonies that had been laid down for man's use by supernatural beings far back in the past—for example, by Masau (the god of death and also of regeneration), who came and taught the Hopi people

> . . . many things concerning growth of plants and trees and instructed them about planting beans when the moon should be at a certain age and after the sun had come a certain distance on his way back to the north. Many, many days this has been the custom and we have no right to forsake the ways of our fathers. [39]

In other words, the Hopi concern for accuracy was rooted in their respect for the correct performance of ancient rituals, not in a quest for new knowledge.

The ingenious study of the Hopi calendar published in 1977 by Stephen McCluskey, historian of science at West Virginia University, demonstrates how much their scheme differed from our own. There were complicated rules for beginning each major ceremony, some connected just with the sun, others involving both lunar and solar counting procedures. They were all short-term calculations; the priests seldom had to count more than sixteen days from one of the solstices or from the appearance of a particular moon. Apparently, the Hopi did not attempt to unify the calendar as an orderly whole. For this reason, the disparity between the rhythms of the sun and the moon naturally created confusion from time to time. Any doubts seem to have been resolved at the start of each annual sequence of ceremonies around November, when the lunar months were adjusted so that an extra month could be slipped into the sequence if necessary.

The Hopi understanding of these calendrical problems was not at all like our own. McCluskey states that when they encountered difficulties in timing ceremonies, the sun or moon would probably be thought of as "running fast" or "running slow." The idea that celestial objects followed regular, uniform motions through the sky—so vital to our own astronomical concepts—did not enter the Hopi mind.

Even the basic arrangement of the Pueblo year strikes an unfamiliar note. In 1979 the Pueblo of Zuni issued a wall calendar, which appears to be a marriage of convenience between European time intervals and traditional names and concepts. While the months and their arrangement are no different from an ordinary calendar, their names follow a strange pattern. The first six months bear titles that relate sensibly to seasonal occurrences,

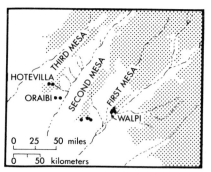

Map of the Hopi villages, with names of those mentioned in the text.

such as "Trees Broken by Snow Moon" (January) and "No Snow in Trails Moon" (February, anticipating the spring thaw), but the self-same month names also apply to the second half of the year from July to December. In the 1979 calendar, an effort is made to explain this away by the suggestion that the second "Trees Broken by Snow Moon" really refers to ripe fruits breaking the branches of trees in July, although how snow can possibly thaw in the trails in August passes without comment.

A century ago, the Zuni calendar was probably less confusing, for Cushing and others claimed that the months from July to December were in fact named after colors (the special colors associated with the six sacred directions of Zuni space). In general, the splitting of the year served to separate the activities of preparation and planting from those of fruition and harvesting; the season of agricultural work was divided from that of hunting, gaming, warfare, and curing.

The Hopi year of thirteen months was also split into two repeating cycles. The most important ceremonies were staged during the seven winter and spring months. The month names were then repeated for the remaining six summer and autumn months, as in the modern Zuni calendar described above. Quite possibly, the Zuni once recognized thirteen months, too, for their present total of twelve may reflect an old adjustment to European habits.

Even though they have come down to us in a confused form, the Zuni and Hopi calendars reflect something of the passion for intricate detail, order, and regularity apparent in so many aspects of Pueblo life. These concerns, however, had little to do with the logic of our own calendar.

The division of the year into repeating halves was more than a mere eccentricity on the part of the Hopi and Zuni. The significance of opposites preoccupied them in other matters besides that of the calendar. For instance, all Pueblo people recognized paired beings and incidents in their mythology, notably the powerful twin War Gods who were often represented as the Morning Star and Evening Star.

By day, the Sun Father of the Pueblos traveled over the world of the living, while at night he crossed and illuminated the underworld of the dead. There, time and space were "opposite" to their living counterparts. A popular Pueblo folktale concerns a girl who, revived from the dead by witches, could only work in the night and sleep during the day. The moons and seasons could be thought of as revolving on a great wheel, so that the spirits below experienced winter during the "real" summer, and so on. Even the underworld corn harvest would be repeated exactly on earth.

The vision of time cycles among the southwestern tribes could lead to the idea that their experience of past and present is somehow quite foreign to our own. For many years the influential

THE YEAR OF THE HOPI: "BOUND UP IN TIME"

American linguist Benjamin Lee Whorf claimed that the Hopi were unable to express lengths of time or space in the way we do. Because of this peculiarity, they could not speak or think directly of intervals between events. In other words, the difference between "us" and "them" is deeply embedded in language itself and in the working of the mind. Whorf's conclusion that the Hopi cannot refer easily to time intervals is surprising if we consider the skill that all Pueblo peoples displayed in calendar reckoning and sunwatching.

Doubts about these theories were confirmed when the German linguist Helmut Gipper conducted fresh studies among the Hopi during the late 1960s. His work shows that the mental landscape of the Hopi is not really such an unfamiliar one. Their shortest unit of time is the day; they have no words equivalent to our hours, minutes, or seconds. However, time and space intervals can certainly be expressed in ordinary speech, since they count days and nights between ceremonies and talk about them in a straightforward way.

Yet with no regard for clocks, and lacking a sense of recorded history, the Hopi do tend to think of past and present as one and to treat the future as something separate to come. According to Gipper, they have not yet developed the detachment from events that we know: "they live in time but not *apart* from it, they are bound up in time but are not neutral observers of objective physical time."[40] He concludes that everyday speech, thought, and outlook are affected by the seasonal pattern of their lives and by their belief in a universe of endless cycles: day and night, summer and winter, the living and the dead.

Similarly, the Hopi use language just as we do to refer to length, breadth, and direction, and yet here the special character of the Pueblo outlook is even more apparent. Supernatural power is not exercised at random in the Pueblo world, but is controlled and received from six sacred directions of space. The pueblo lies at the middle point of these six influences, which take the forms of

Hopi Snake Priests setting out from their kiva *at the village of Oraibi toward the north on the first morning of the four-day hunt for snakes to be used in the Snake Dance. Photographed in August, 1900.*

colors, predatory beasts, birds, sacred houses, even special lakes and trees, each associated with a particular direction. These beings and objects loom large in folktales and myths, and are carefully taken into account in the procedure of every ceremony. For example, the Hopi still reverently observe the sequence of directions in their celebrated Snake Dance, when dancers clasp live rattlesnakes in their arms or mouths and always pass counter-clockwise around the plaza. Before the ceremony begins, the snakes are hunted among the barren outcrops on four separate days, each day from a different snake shrine located to the northwest, southwest, southeast, and northeast of the village.

Two of the sacred directions are "up" and "down": the zenith vertically overhead and the nadir directly underground. The other four correspond not to the cardinal points familiar to us but to the positions of sunrise and sunset at midsummer and midwinter. So the solstices not only form the turning point of the ceremonial calendar, but also define the directions of space. The Hopi are influenced by watching the sun's movement on the horizon as deeply as are the Zuni.

Since both tribes share many ceremonial traditions, it is not surprising that about a century ago the duties of the Hopi sunwatchers were quite similar to those of the Zuni *Pekwin*. For example, they were responsible for timing the critical "calling back of the sun" from his southwestern house at midwinter. At the village of Walpi, perched high on a mesa top overlooking the level desert, the Sun Priest installed himself on the roof of the Sun Clan house and watched each winter evening as the sun set among the distant peaks of the San Francisco mountains. When it eventually reached a dip at the far edge of the mountains, it was time for the Sun Priest's announcement (made by the "Town Crier") proclaiming the start of the night-long solstice ceremonies. Then the initiates crowded into the gloomy *kivas* to watch

The Hopi village of Walpi, dramatically situated on a narrow mesa overlooking the Painted Desert. This view, taken by W. H. Jackson in 1872, has not changed substantially since then.

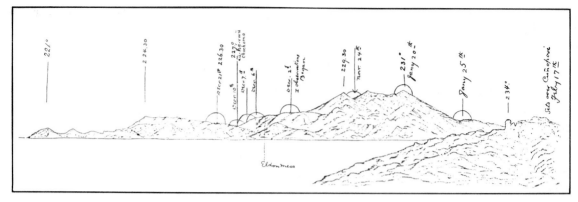

A Hopi horizon calendar. The anthropologist Alexander M. Stephen, who lived among the Hopi from 1891 to 1894, recorded a series of solar observations made by the Sun Priest for the timing of midwinter ceremonies. The horizon is the San Francisco Peaks near Flagstaff, viewed from the roof of the Bear Clan house at Walpi.

ritual dances and dramas that must have rivalled those of the Zuni *Shalako* event. The awe-inspiring atmosphere was enhanced by such devices as a great screen that concealed the actors and displayed cloth-covered effigies of serpents. These snakes appeared to weave and twist in the air, controlled by almost-invisible horsehair strings.

Unlike the *Pekwin* of Zuni, the Hopi sunwatchers were expected to time other events besides the major solstice celebrations. They included hunts, marriages, and, perhaps most significant of all, the dates on which crops were to be planted. At several Hopi villages, early anthropologists noted the existence of horizon calendars memorized by the Sun Priest. The calendars consisted of up to twenty dips and notches on the distant skyline. Each one was clearly identified and marked the signal for the beginning of a particular crop-planting operation during the agricultural season.

HOPI PRIESTS AND PLANTERS

Why were the Hopi so preoccupied with sunwatching? Observing conditions were certainly favorable, for the Arizona desert sky is often brilliantly clear at sunset, while eroded sandstone pinnacles and far-off snowy peaks form striking outlines around the Hopi horizons. However, the Hopi's intense concern with ordering time and space is a remarkable aspect shared by many other southwestern societies and requiring more than good visibility as an explanation.

There are some, of course, who feel that the entire pattern of religious thoughts and observances among these peoples is inscrutable and beyond our understanding; in the view of one writer, Hamilton Tyler, it fulfills "psychic needs," and there is little more to be said about it:

> The function of Pueblo religion is psychic. You can clean your field after winter abates without rites, and you can hoe your corn because you have to; but there is undoubtedly more satisfaction, and ease, in tending the crops with rites as well as a hoe.[41]

Attending a ceremony at a Hopi village, one cannot fail to be struck by the contrast between the vigor and complexity of the dances—the brilliant colors and resonant chants—and the absolute stillness and emptiness of the surrounding landscape. So arises the popular idea that intricate ritual beliefs are a compensation or psychological support for the hardness of life in an unyielding environment. The solitude of the desert is less forbidding if it is peopled with dozens of supernatural beings. In the words of Hamilton Tyler, the purpose of Pueblo religion "is to enrich a sparse way of life, to establish psychic satisfactions where material ones are hard to come by."[42] Whether or not we agree with this point of view, a more obvious possibility exists that we can relate elaborate ceremonies and accurate sunwatching to basic problems of survival in the desert.

Growing such staples as corn and beans in the Hopi region has always been a risky business. The average rainfall of ten to thirteen inches is too low to support crops without special methods of cultivation. The rains occur mainly during scattered, often violent, thunderstorms in July and August. A downpour may bury one crop under an avalanche of mud, while another field not far away stands parched and dry. There are no permanent streams to permit irrigation on a big scale, although after a storm the volume of water that races destructively along the valley beds can equal the flow of some of the largest American rivers.

Channeling the water from flash floods and runoffs is a special skill that the Hopi learned over the centuries. Their fields were carefully positioned near normally dry stream beds, and much of their effort was directed toward ensuring that these plots were inundated but not swept away. Constant attention to the upkeep of ditches and low dams was essential, particularly in the midst of the storm itself. As an observer noted half a century ago, "an everyday sight during showers is the irrigator with hoe or stick, or even with his hands, constructing ridges of earth or laying down sagebursh in such a manner as to insure a thorough soaking of his planted field."[43] The fields owned by any single Hopi family group or clan were widely scattered all over the landscape, taking advantage of patches of favorable drainage wherever they could be found. The risk of crop failure was somewhat offset by the different rates of corn growth resulting from these varied locations. In addition, extra insurance was provided by small plots watered by other methods, such as crops tended on the less arid sand dunes high up on the mesa tops. All these methods were laborious; every technique demanded experience and luck.

Once corn and bean rows were established, there was the threat of severe frost, which might occur as late as June or as early as September. Since corn matured slowly under such conditions, the average frost-free period of about 130 days was scarcely longer than the growing season of the plant itself.

Once again the Hopi devised an "insurance policy" to guard against the possibility that crops would be wiped out by frost. They scattered the planting dates widely through the early summer season (each clan beginning its cultivation at a different

A Hopi farmer, photographed around the turn of the century by Adam Clark Vroman.

time), and so it was less likely that all would fail. The cycle began cautiously around mid-April with a few well-sheltered plots sown for the "green corn" ritual at the end of July. The main plantings fell between mid-May and the summer solstice, resulting in harvests toward the end of September, by which time frosts would already have occurred.

In this unpredictable situation, a fixed schedule of planting dates, which reflected the experience of generations of farmers, was obviously useful to the community. The most directly practical information provided by the sunwatcher was a limit of a few days after the summer solstice beyond which plantings could not be expected to survive the onset of winter. From the records of anthropologists who stayed with the Hopi for more than a single year, we know that the sunwatchers were capable of real precision, the dates for planting varying by not more than a day or two from one year to the next. Time intervals of only four or five days between some plantings also suggest remarkably precise observations by the sunwatchers.

Their skills were not carefully guarded secrets, but were common knowledge among the community. The important fertility rites of the *Powamu* ceremony, staged in the underground *kivas* a few months before plantings actually began, made the connection between sunwatching and the successful fruition of the crops clear to all the participants. During *Powamu*, the rising and setting points of the sun along the horizon were solemnly chanted aloud. There were also "smoke talks" preceding the announcement of each ceremony, providing an opportunity for anyone familiar with traditional lore to criticize the sunwatcher's performance. The moral purity of the sunwatcher was thought to be

To master their unpredictable environment, the Hopi exploit many different locations and techniques, including modern irrigation devices, above, and terraced fields, below.

no less important for the well-being of the Hopi community at Walpi than at Zuni, as an anthropologist once recorded:

> We think the sunwatcher is not a very good man. He missed some places, he was wrong last year. And I think he is going to miss or take on some place [i.e. observe inaccurately]. All the people think that is why we had so much cold this winter and no snow.[44]

The abilities of sunwatchers were so widely discussed that they were even commented on by people in other Hopi villages; at Hotevilla, the skipping of a month in 1933 was said to have provoked "much teasing" from their neighbors.

Furthermore, there was no strict enforcement of the instructions given by these officials. At Walpi a certain formality attended the Town Crier's announcement of the first planting, for this was due to be carried out by everyone for the benefit of the village chief. By contrast, the sunwatcher at Oraibi simply spoke to his neighbors, who spread the word around among the families. The announcements were often ignored by the individual farmer, especially when a sunwatcher became confused in his observations. As the anthropologist Mischa Titiev explains, "On the whole there was no compulsion regarding the planting dates for various crops but it was felt best for people to adhere to the schedule which their ancestors had worked out."[45] Imposed by tradition rather than by authority, the Hopi horizon calendars must have introduced a measure of reliability into the uncertain hazards of desert cultivation.

So, too, many aspects of southwestern religion and society that "put the individual in his place" in a firmly ordered world must have been important for survival in the long run. Personal identity was lost in the splendid formality of public dances with masks, costumes, massed chants and dance steps. A highly self-disciplined society had an advantage in an environment where lean times were to be expected and precisely timed "hard labor" in the fields was essential for existence. In this light, even the most bizarre ceremonies "make sense" to us in very general terms. The intricacy and elaborate repetition of the rituals each year encouraged the solidarity of the Pueblo in an insecure landscape.

In the minds of the Hopi or Zuni, of course, explanations were simpler: the ceremonies were there to do the job for which they were intended. A well-staged ritual actually *could* extend the growing season by days or weeks. When the *Pekwin* trekked out to the sun shrine at Matsakia at the summer solstice, his mission helped to restore a sense of order in the community by the accurate setting of the calendar. But for him, the purpose of his visit was quite literally "to make the sun come out." A narrative by an elderly Zuni recorded in the mid-1960s reminds us of this belief. A woman tries to seduce the Sun Priest, arguing,

> "He's going to come out ANYWAY.
> Why don't we go to my field" she told him.
> "Well I won't go.

The Powamu ceremony is one of the most vital Hopi rituals, since it is believed to help secure successful harvests. During Powamu, the Heheya Katchinas of Walpi, photographed here in 1893, pretend to copulate with spectators.

I didn't come just to go anywhere, but to bring out my Sun
        Father."
That's what he said. "Indeed.
But he's going to come up ANYWAY
Just the way he's BEEN coming up" that's what
She said. "No, it's because of me that he comes up."
That's what he told her. . . . [46]

Symbolic designs copied from sand paint-
ings used for healing rituals by priests at
Acoma pueblo, New Mexico.

And he is proved right after she murders him, for the sun stays
down. For us, this idea makes no more sense than the Hopi
notion that the power of the sun could replace a lost tooth. Both
ideas do, however, express faith in a precise link between human
actions and natural forces, and in general terms we can understand the significance of this to the insecure desert farmer.

Our "common sense" explanations, of course, tell us little about
how beliefs and ceremonies arose in the first place. How did the
original horizon calendar come about, and how old are such
practices of solar observation in the Southwest? Is it possible that
the sunwatchers once had more power and were more faithfully
obeyed than was the case in recent times? With the help of
anthropologists, we can make our crude attempts to understand
the motives and logic underlying ceremonies, but the "how" and
"when" can rarely be guessed.

In the Southwest, however, clues *do* exist to the way of life
followed by the ancestors of the modern Pueblo peoples. The
famous cliff dwellings of Mesa Verde and the multistoried "apartment blocks" of Chaco Canyon are among the most impressive
prehistoric remains in North America, and were obviously the
work of a highly developed culture. At these and other sites,
mysteriously abandoned over six hundred years ago, the ruins of
kivas suggest that the religious practices there may have resembled the traditions of the modern Pueblo.

In fact, the possibility that the present-day ceremonies of Zuni
and Hopi preserve fragments of ancient beliefs has always intrigued scholars of the Southwest. It was therefore particularly
exciting when the first convincing examples of solar alignments
and probable sunwatchers' shrines were identified at ancient sites
in that region during the 1970s. This was the work of several
pioneering researchers, including the astronomer Ray Williamson and the anthropologist Jonathan Reyman. As explained in
the next chapter, their findings added new weight to comparisons
between the world of the living Pueblo and that of the vanished
cultures.

The observatories of the ancient sun priests stand unroofed and
empty, but the rooms in which they waited for the dawn are still
lit by the first rays of the sun as it shines through the narrow
windows today.

11. *The Astronomy of the "Old Ones"*

In the summer of 1972, an archaeological survey team from the University of New Mexico discovered a remarkable painting on the western wall of Chaco Canyon, a remote sandstone valley about one hundred miles northwest of Albuquerque. Here, on the undersurface of a shallow projecting ledge about twenty feet above the ground, they found a striking group of painted symbols: a star, the crescent moon, and a hand. Also, on the vertical cliff wall just beneath, there was a symbol recognized by the modern Pueblo as standing for the sun: a dot encircled by two rings. These paintings are a short walk from the ruins of Penasco Blanco, the name given to a settlement now reduced to stumps of crumbling stone. A thousand years ago, between 900 and 1100 A.D., Penasco Blanco was a splendid crescent-shaped block of "apartments," which may have reached as high as three stories.

The interpretation of ancient sky symbols carved or painted on rocks poses special difficulties. It is rarely possible to date such rock art with confidence, and the meaning of the astronomical symbolism is not always clear. Thus Cushing's mention of carvings associated with the Zuni "observatory" at Matsakia is particularly interesting, even though the site and the markings he described have disappeared. We may recall that the *Pekwin* "approached a square open tower and seated himself just inside upon a rude, ancient stone chair, and before a pillar sculpted with the face of the sun, the sacred hand, the morning star and the new moon. . . ."[47] If a sunwatcher's shrine was decorated with this collection of symbols only a century ago, could similar markings from a much earlier period be used to identify the site of an ancient observatory?

The painted signs at Chaco Canyon offer a startlingly close parallel to Cushing's account of Matsakia. Did they mark the sacred spot where the tenth- and eleventh-century sun priests of Penasco Blanco stood to watch the sunrise? Ray Williamson, a leading investigator of prehistoric astronomy in the Southwest, thinks that the priests may have climbed to the top of the rock face immediately above the painting. At this spot, close to an ancient roadway, they could have watched the winter sun as it rose in line with the edge of a prominent cliff interrupting the local horizon. Williamson's suggestion is made more convincing by his identification of another large painting of a sun, or star, this time executed in white and located near the ruin of Wijiji at the other end of the canyon. This panel, too, seems well placed for observations of the winter solstice.

If these ideas are correct, then it would seem that the inhabitants of Chaco Canyon a thousand years ago had officials similar to the recent Pueblo sun priests who watched the horizon for the

THE STAR,
THE MOON, AND
THE SACRED HAND

auspicious timing of rites and ceremonies. Could the painted symbols, however, mean something else quite unexpected?

The discovery of the Penasco Blanco painting was an event of significance outside the relatively small circle of people interested in the archaeology of the Southwest. For many years, astronomers have wondered if ancient historical records might conceal important clues to the nature and timing of spectacular events in the sky. The tremendously violent explosions of stars known as supernovae are of particular interest, for a better understanding of these phenomena might lead to general insights about the processes by which stars evolve and decay.

The best-known and most carefully scrutinized of all these exploding stars is the Crab Nebula in the constellation of Taurus. The Crab is a strikingly beautiful cloud of gas expanding from its

An imaginary recreation of the 1054 A.D. supernova close to the crescent moon, and the cliff painting at Penasco Blanco in Chaco Canyon that may depict the event. Photomontage by Smithsonian astronomer Von Del Chamberlain.

source at the speed of 1,300 kilometers per second. What this object looked like at the time of its birth nine hundred years ago is a matter of conjecture. The most widely accepted estimate credits the supernova with five or six times the brightness of Venus, itself the most conspicuous object in the sky other than the sun and moon. Such a brilliant gleam in the heavens would have been visible even in daylight for a period of some three weeks after the first explosion. Even two years afterwards, it would still have been a notable night-sky attraction.

Scientists concerned with the Crab Nebula have noted a curious phenomenon, the seeming lack of eye-witness reports of its sudden appearance. So far as is known, no one in eleventh-century Europe saw fit to jot down even a passing comment about this dazzling new star, although a probable reference in one Arab chronicle has recently been uncovered. Only the observant astronomers of China and Japan seem to have noted the arrival of the "guest star," set down in their annals during the first week of July 1054.

The knowledge that people in the American Southwest occupied settlements such as those in Chaco Canyon during the mid-eleventh century has proved an irresistible lure for the supernova investigators. From this part of America, the first appearance of the Crab in the dawn sky may have looked particularly impressive because it was close to the waning crescent of the moon—a mere two degrees to the north of it. Any depiction of this event on a carved or painted rock surface should therefore include a prominent star symbol arranged in proper relationship to the sign of the crescent moon.

The first two cases of such a motif in the rock art of Arizona were spotted in 1955 by the astronomer William C. Miller. Since

Crescent-and-star carving located near the ruins of Village of the Great Kivas, in the Zuni Reservation, New Mexico.

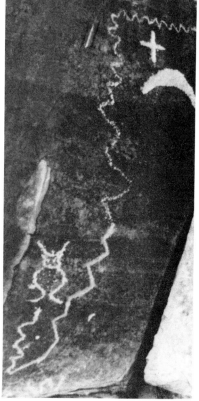

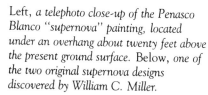

Left, a telephoto close-up of the Penasco Blanco "supernova" painting, located under an overhang about twenty feet above the present ground surface. Below, one of the two original supernova designs discovered by William C. Miller.

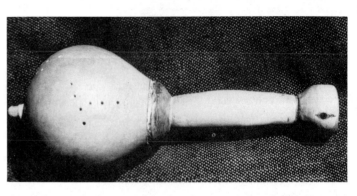

*Two views of a Navaho gourd rattle,
showing holes drilled to resemble part
of the constellation of Orion.*

then, the total of reported star-and-crescent designs has risen to more than twenty, scattered widely across the Southwest from northern California to Texas. If these prehistoric paintings and carvings do refer to the supernova explosion, even indirectly as a legendary event, then it is fair to conclude along with Ed Krupp, author of *In Search of Ancient Astronomies*, "that these early Americans knew the sky well enough to recognize a newcomer in the field of stars and that, in a limited sense, they knew the sky as well as the more culturally advanced Chinese astronomers."[48]

Actually, the capability of native southwesterners to observe "the field of stars" accurately is not to be doubted. Until recently, the Navaho embellished their sacred sand paintings and ceremonial gourd rattles with star signs. These were often grouped in recognizable patterns, corresponding to constellations such as the Pleiades, parts of Orion, Scorpio, and so on, although naturally the Navaho used their own traditional names to refer to them. It is certainly reasonable to assume that the canyon dwellers nine hundred years ago were capable of noticing and depicting a specific event in the sky. The question is: did they do it?

The painting at Penasco Blanco does show a "star" at the proper angle in relationship to the crescent, which is also facing the correct way round, just as the supernova and the moon actually appeared one morning in early July of 1054. However, it can be argued that the habit of recording historical events is characteristic of a European rather than a Native American outlook.

Furthermore, according to Cushing, the star sign engraved on the Matsakia shrine represented Venus as the Morning Star. In fact, starlike Venus symbols were often painted alongside moon and sun designs on the ritual masks, murals, and sacred images of the Pueblo recorded by anthropologists during the past century. As the Morning and Evening Star, the planet was identified with the Twin War Gods, perhaps the most widely venerated figures in Pueblo mythology. These youthful warriors were the allies of the First People, saving them from many fearful monsters. In addition, they aided their father, the sun, in making his daily journey across the sky, and ensured that he returned north again after the winter solstice. Although recent Pueblo lore may be an unreliable guide to understanding a thousand-year-old painting, such associations suggest that the Chaco motifs may have marked a sunwatcher's observatory or shrine, rather than a particular event in the sky.

Of course, we will probably never know exactly what the ancient carved and painted sky symbols of the Southwest represent. The supernova theory is only one possibility among many. But as in the case of the Fairy Stone discussed earlier, the theory shows how eagerly researchers of today tend to impose their modern interests and scientific outlook on ancient cultures.

THE SUN TOWERS OF HOVENWEEP

Since one guess about the meaning of a rock painting is often as good as another, more solid evidence of ancient astronomical observations is clearly essential. Besides discovering the existence of possible sunwatchers' shrines in Chaco Canyon, Williamson and his colleagues also pioneered the study of astronomically aligned architecture in the Southwest.

Some of their most interesting work was carried out at a place that the Indians of the Ute tribe called Hovenweep, the "deserted valley." There could be no better name for such a desolate landscape, reached by miles of dirt road across the rolling mesa of the Four Corners (where Utah, Colorado, New Mexico, and Arizona meet). In this vast wilderness, travelers are surprised to encounter the ruins of small stone towers, some resembling medieval turrets out of a child's picture book. Even the largest and most solidly built structures, dating mainly to the thirteenth century A.D., rarely exceed two stories in height or two meters in diameter. The peculiarity of the towers is heightened by their isolation along lonely gullies draining into the San Juan River, or perched high on pinnacles of rock. Why were so many towers built, each on such a small scale? And who were the inhabitants of this forlorn country? The local Navaho know them as the "Anasazi," a word meaning simply "the old ones." The name reflects the sense of mystery that even native southwesterners felt about their ancestors in the area.

Hovenweep Castle, Utah, resembles a European fortress of the Middle Ages, but it seems to have been built partly as a solar observatory by the Anasazi around 1200 A.D.

Midsummer sunrise viewed through a wall slot at Unit House, one of the many ruins at Hovenweep Canyon. The deliberate arrangement of this slot is suggested by its carefully plastered surface and its unusual angle in relation to the rest of the wall.

The largest of the ruins in Hovenweep Canyon, in the southeastern corner of Utah, certainly does leave a puzzling impression on the visitor. From a superficial inspection of Hovenweep Castle, it would be hard to guess what exactly this rather unsightly, straggling block of rooms at the canyon rim might be. If it was indeed a castle, then the other smaller towers scattered around it seem like eccentric afterthoughts; if it was a family dwelling, the windowless lower rooms must have been dim as a dungeon.

One evening in late June, Ray Williamson led me inside a corner block apparently added onto the western side of the Castle. It was an empty chamber, with bare stone walls, featureless except for three low entrances and two tiny slots opening to the outside air. That evening, close to the midsummer solstice, we watched as the sun began to set, the roof of sky above us turning brilliantly blue while the chamber around us became progressively gloomier. Then, abruptly, the sun appeared framed in one of the little holes, while a patch of yellow light shone for several minutes at one corner of the room, close to the lintel stone of one of the low inner doorways.

It would be easy to dismiss this as a coincidence—except for the fact that *exactly* the same event happens at midwinter. At that time, the sun can be seen setting in the *other* hole, and a square of light falls beside the lintel of the *other* inner doorway in the room. At intervening periods between the solstices, the patches of light would appear somewhere on the wall between the two interior doors; the observers could have marked off a crude calendar on the original wall plaster, which has long since crumbled away.

The position of the third doorway, opening out over the giddy drop of the canyon wall below the Castle, strengthens the astronomical argument further. At the equinoxes, the sun lines up with this third entrance and with one of the interior doors. In fact, all the features of this room seem to have been planned in relation to the movements of the setting sun. What, then, is an astronomical observatory doing on the ground floor of a supposedly defensive "castle"?

Most of the towers were constructed for a variety of purposes, ranging from the ceremonial to the utilitarian. That, at any rate, was the conclusion of archaeologists from San Jose State University, California, who dug test trenches inside seven towers in 1976. Some clearly had a sacred function, since they are located close to *kivas* (in certain cases, actually connected to them by a subterranean tunnel), or else *kiva*-like features are set in the tower floors. The San Jose team discovered corn-grinding stones inside one of the Hovenweep towers; in another structure at nearby Holly Canyon, they found objects probably used in ceremonies, such as a vessel containing the partial skeletons of a rabbit, lizards, a mouse, and a toad. The larger Hovenweep towers apparently had several rooms serving different activities; one housed a *kiva* surrounded by storage or workrooms.

The excavations revealed that the towers were not reserved exclusively for astronomical observations, or in fact for any other single purpose. Indeed, Williamson and his team were able to locate only two other ruins outside the main Hovenweep group which seemed to offer convincing examples of astronomically aligned wall slots. At the Cajon site, the room featuring the wall slots closely resembled the room at Hovenweep Castle. The architecture of the Cajon room shows that it was a later addition to the main tower, just as it was at Hovenweep. So the two cases of specially laid out "solstice rooms" give the impression of being afterthoughts, supplementary to the principal functions of the towers, which were probably used for both sacred and everyday activities.

The tradition of tower building seems to have been neither widespread nor long-lasting. We do not encounter towers built more recently than the end of the classic third phase of Pueblo construction, around 1300 A.D., and the fashion seems to have been restricted mainly to the Four Corners region. So we cannot automatically assume that there was a connection between the activities at Hovenweep and later practices, such as those of the *Pekwin* who waited for sunrise in the Zuni Sun Tower, or the *Cacique* of Cochiti who studied the eastern skyline through the special window in his office.

Although I was highly impressed by the experience of witnessing the setting sun framed by the wall slot, the case of Hovenweep Castle illustrates some typical problems encountered by investigators in the Southwest. It is difficult to assess the astronomical theories when the structures were clearly built for a variety of purposes, and when the link with traditions of the modern Pueblo is uncertain.

Fortunately, recent discoveries in Chaco Canyon shed further light on these problems. The new finds increase the likelihood that the Anasazi *did* engage in sunwatching activities similar to those of the modern sun priests.

Ruined tower along the McElmo valley in the Four Corners region.

DAGGERS OF LIGHT: THE SUN SHRINE DISCOVERIES

Of all the ruins of the Southwest, none are more haunting than the thirteen great "towns" which lie abandoned in the wastes of Chaco Canyon. This impression is due partly to the isolation of the canyon and the effort required to reach it across twenty-five miles of bone-shaking dirt road. Distance and space slowly lose their meaning across a rolling emptiness of sagebrush, grass, sand drifts, and dry steambeds. Finally, after arriving on the broad avenue of the canyon floor—at some points nearly a mile across and bound by sandstone walls two hundred feet high—a sensation of vastness remains with the visitor.

This impression forms a contrast to the compact human scale of each of the Chaco towns. Inside these ruins, living space is neatly divided into compartments. The orderly tiers range two, three, or more stories up toward the boundless sky. On the rim of the great walled settlements known as Pueblo Bonito and Chetro Ketl, the rooftop view of the inhabitants must have taken in the whole community and the activities of everyone on the adjoining roofs and in the central plaza below. Even today such a viewpoint, gained precariously among the crumbling walls and reconstructed floors, conveys at once the purposeful planning and earnest community sense of a long-vanished people. Here one still feels something of the timeless discipline that governs the living Pueblo today. Here, also, is a sense of defeat. The empty doorways gape over thick sagebrush and juniper, now fit only for lizards and rattlesnakes; the dry river channel cuts deeply and uselessly into the dead earth.

It is hard to picture this landscape as it was less than a thousand years ago. There were roads crisscrossing the canyon, busy with farmers and traders. The roads ran from the great towns to gullies in the cliff walls where stairways were hewn laboriously to ease the traffic. Along these routes, consignments of timber and high-quality stone came from the surrounding mesa, hand-pulled or pushed, for there were no draft animals in the canyon. Luxury goods of shell, turquoise, copper, and other prized materials arrived from farther afield. Ingenious systems of ditches and check dams were positioned close to the canyon walls to divert the runoff water from summer storms into canals and reservoirs. There were gardens dotting the canyon floor, green plots carefully laid out in square grids and watered from the canals or from the shallow riverbank.

Where there is silence and sterility today, an estimated six

View from the cliffs overlooking Pueblo Bonito, Chaco Canyon.

thousand people were once supported in numerous settlements. These included towns that were elaborately planned by some type of authority within the community. This organizing power probably also expressed itself in splendid ceremonies, which bound the people together. The ceremonies took place in *kivas* resembling big sunken amphitheaters up to fifty feet across; there are more than thirty of them (not all contemporary) at Pueblo Bonito alone. The priests were perhaps adorned with turquoise beads and masks of painted wood; the air rang with fanfares of conch-shell trumpets.

Pueblo Bonito remains the most awe-inspiring of the Chaco towns, particularly the splendid sweep of its curving north wall nearly five stories tall. The stonework is an artful blend of fine and coarse textures, and shows that the builders paid close attention to details of architectural style. About eight hundred rooms were constructed against this wall, grouped in tiers around the rim of the crescent so that they encircled the *kivas* and the broad open space of the plaza.

Estimates of the total population once present at Pueblo Bonito have fluctuated wildly since the first excavators dug there in 1897. A recent survey suggests that five hundred may be near the mark. The direction of communal energies on this scale surely indicates that special social forces were at work—a vital sense of collective or tribal identity, perhaps, or else a strong coercive power.

The authority of the Chaco town planners was reflected in their highly organized use of space. For example, the layout of Pueblo

A typical example of the fine masonry incorporated in Pueblo Bonito.

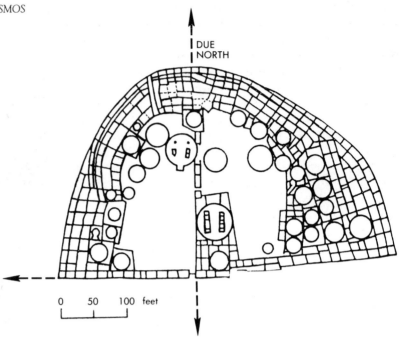

Plan of Pueblo Bonito, showing the alignment of two of the principal walls to the cardinal directions.

Plan of Casa Rinconada, showing the major ground features and the twenty-eight wall niches, all arranged according to the four directions. The kiva is nearly 20 meters in diameter.

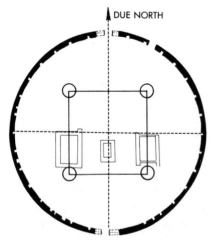

Bonito seems to have been based on the four cardinal directions familiar to us (unlike the four solstice directions emphasized in recent Hopi tradition). The straight southwest wall is aligned due east–west; a north–south wall neatly divides the central plaza in two. Next to this wall lies the largest of all the Bonito *kivas* (some fifty-two feet in diameter). The arrangement of the single entrance and of the wall and floor features of the *kiva* shows that it, too, was set out on a north–south plan.

Why was this careful ordering of space within the underground sanctuary important? According to modern Pueblo beliefs, the *kiva* symbolizes the place where the first humans emerged from the lower world. The traditions of Acoma Pueblo, New Mexico, assert that the first *kivas* were circular, even though the present-day examples are all rectangular. One Acoma informant interviewed in 1928 explained how supernatural ancestors had decreed the layout of the *kiva* so that it would be a sacred model of the world. The four central pillars supporting the roof were to represent the trees planted in the underworld for the first humans to climb. Each pillar was associated with a different sacred direction and a special color. "The walls represent the sky, the beams of the roof (made of wood of the first four trees) represent the Milky Way, *wakaiyanistiaw'tsa* (way-above-earth beam). The sky looks like a circle, hence the round shape of the kiva."[49]

We cannot be sure that the Chaco people thought of their huge circular *kivas* in identical terms. However, the largest and most impressive of all the Chaco *kivas*, known as Casa Rinconada, was certainly constructed with careful attention to its geometry and to the four directions of space. The importance of this sanctuary

is evident from its great size (nearly sixty-four feet in diameter) and its setting at the top of a low hill on the south side of the canyon, in dramatic isolation from any of the major towns.

Ray Williamson and his colleague Howard Fisher undertook a precise survey of Casa Rinconada during the mid-1970s. Although the monument was restored by the National Park Service, the recent measurements correspond closely with those of the original excavators in 1931, so we can be confident that the ground plan is authentic. The survey reveals that the *kiva* was laid out along a major axis running north–south. An east–west axis passes through two of the twenty-eight little niches set into the encircling inside wall. These niches are spaced symmetrically and at regular intervals around the axis lines. Another set of cardinal directions is defined by the masonry foundations of the four huge pillars that once supported the roof; the pillars stood at the corners of a square oriented to the four directions. The geometric centers of all these separate features coincide so closely that there must have been skillful control over the planning and execution of the architectural design. This design was probably intended to be more than merely decorative or pleasing to the eye, since, according to the account from Acoma, the layout of the *kiva* stood for the supernatural ordering of the universe itself.

The precision of the plan of Casa Rinconada raises the question of whether a unit of measurement was employed by the Chaco builders. At Pueblo Bonito, about one hundred rooms belonging to four separate construction phases were excavated under the auspices of the National Geographic Society during the 1920s. In the course of this work, hundreds of building dimensions were recorded. In 1972 the anthropologist Travis Hudson published a

Casa Rinconada, most impressive of the kivas built by the Anasazi in Chaco Canyon during the 11th century A.D.

Two leading investigators of southwestern astronomy, Stephen McCluskey, left, and Ray Williamson, debate the significance of Casa Rinconada during a visit there in 1979.

study in which he examined these figures with the aid of statistical techniques. He concluded that standard units of length had indeed been employed over several episodes of construction during the eleventh century A.D. But the same measuring unit was not repeated from one episode to the next. Apparently the architects devised a new unit each time they organized a major building effort.

The most interesting of all Hudson's findings, however, was that during one building phase, *two separate units* seem to have been in use at the same time. The boundary between the two sets of measurements was the central dividing wall that ran north–south. To the west of it the builders used a unit of about twenty inches; to the east, twenty-nine inches. Presumably, two distinct construction teams or planning authorities were responsible for these different practices. The two groups must have cooperated, for the structures built next to the central wall incorporated *both* units in their layout.

This finding suggests that during part of the eleventh century, the people of Pueblo Bonito may have been divided into two social groups, or "moieties." Among many pueblos in the Rio Grande Valley today, moieties are essential to the organization of practical and spiritual affairs. Everyone in the pueblo belongs to one or the other of the two groups. The members of a particular moiety attend their own rituals in their special *kiva;* their responsibilities to the community are distinct, and in some towns they obey the directions of their own co-chief, who governs the pueblo jointly with the head of the other group. Membership in one moiety or the other is not determined by family allegiances, but often depends on whether one's house is located to the north or south of the town center. At Zia, New Mexico, "those who live north of an imaginary east-and-west line, drawn through the village between the north and south plazas, belong to the Wren kiva; those who live south of this belong to Turquoise. . . ."[50] The prominence of north–south and east–west lines in the architecture of Chaco does suggest that the four directions had some important social meaning similar to this modern custom.

Since the layout of the Chaco ruins hints at parallels with modern Pueblo ideas of space, is it possible to detect the remains of special sunwatching structures that may have resembled the Zuni Sun Tower or the Matsakia sun shrine? Alignments to the solstices have been claimed from several unusually placed windows or doorways at both Pueblo Bonito and Casa Rinconada, but in each case the ruined or restored state of the structures casts doubt on the evidence. For a convincing demonstration of Anasazi sunwatching techniques, we must turn to a unique and exciting discovery made in the summer of 1977.

Anna Sofaer is a young Washington artist who, with no previous background in archaeology, decided to attend a summer school devoted to the recording of the countless rock carvings and paintings in Chaco Canyon. In pursuit of this activity, she

boldly climbed the massive rock outcrop called Fajada Butte, which stands like a natural fortress guarding the broad southern entrance of the canyon. The tricky climb to the top (about four hundred feet above the valley floor) involves an awkward movement past a ledge where rattlesnakes habitually sun themselves. The summit could not have been attractive for settlement, since there is not a drop of water or an inch of soil. But broken pot fragments, numerous rock carvings, and the ruins of walls belonging to several small structures all prove that the lonely peak drew the Anasazi there for some special purpose. From the top there is an uninterrupted view of the distant desert horizon.

On a narrow ledge close to the summit, Anna Sofaer made her discovery. As she walked along the ledge, her eye was caught by three sandstone slabs stacked vertically on end, one behind the other, all leaning up against the rock face at the same curious

Fajada Butte, a natural outcrop rising about 400 feet above the southern entrance to Chaco Canyon.

These three strangely angled rocks are located on a ledge near the Butte's summit.

angle. This unnatural placing of the boulders aroused her curiosity. Peering between the slabs, she could see two spirals of different sizes carved on the cliff surface just behind them.

These fairly common symbols would not normally have caught her attention. When she returned to the summit the next day, however, she witnessed a remarkable lighting effect. It was her good fortune to be there a little before noon only a week after the summer solstice. At this season, the arrangement of the boulders (to be precise, the narrow gap between the left and middle stones) causes a dramatic play of sunlight across the carved surfaces. A spot of light appears on the outer top edge of the large spiral and quickly grows to a long knifelike sliver of brightness, passing vertically right through the center of the spiral. Just eighteen minutes after its first appearance, the light form shrinks to a spot once again near the opposite edge of the spiral from where it began.

The performance of this "dagger of light" at once impressed Sofaer, as it has every subsequent witness of the event. The perfect division of the spiral by the sun is certainly remarkable, but could this striking combination of the rocks and the sunlight just be a matter of chance?

With the help of her colleagues, Sofaer set out to show that the lighting effect she had observed was a deliberate contrivance. At the winter solstice—and only within one week of that date—the same large spiral is exactly framed by *two* long slivers of light cast by *both* gaps between the three stones. The spiral stands "empty of brightness" at this dead season of the year, but when the sun is at its hottest in midsummer, the center of the carving is filled with light.

The Anasazi "sun dagger." The large spiral carved at Fajada Butte, seen here at midsummer bisected by a "dagger" cast by the noon sunlight.

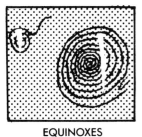
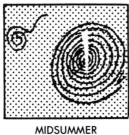
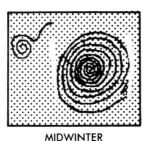

EQUINOXES   MIDSUMMER   MIDWINTER

The correspondence between the carvings and the seasons did not end there. At the equinoxes, a little vertical dagger crosses the center of the *other* smaller spiral. Only a week before or after the equinoxes, the light will have moved distinctly to one side or the other, away from the middle of this little spiral.

Sofaer and her associates think that the Anasazi could have used the carvings as a calendrical device by patiently observing the changing patterns of light. It would have been possible for them to detect the true date of midsummer to within four days, and those of midwinter and the equinoxes to within a week. In theory, if they had watched the shadows cast by moonlight as attentively as those cast by the sun, observers could also have followed the long-term cycle of the moon. The really striking evidence, though, is the effect of the sun. It is difficult to believe that so exact a link between the shapes of the carvings and the light forms thrown at the critical seasons of the year could be due merely to chance.

But how was this remarkable configuration of boulders—in reality, a sensitive indicator of the changing sky—set up in the first place? Sofaer and her colleagues think that the Anasazi deliberately maneuvered the three slabs into position. If their theory is correct, this would represent a remarkable feat, because a simple set of vertical-side slabs would not work. To achieve the straightness and smooth progression of the midsummer light down the spiral, the shape of the slit presented to the sun actually had to be *curved*. We can imagine a lengthy process of trial and error before the stones were placed at exactly the right angle to create the desired outlines on the cliff behind. After that, the curved edges of the slabs where the light passes through may have been artificially shaped to improve the appearance of the "dagger" still further.

On the other hand, an examination of the site in 1982 by an independent team of geologists and archaeologists disagreed with this explanation of the site. In their view, all three slabs originated in the horizontal rock layers immediately above and to the left of their present position. The slabs were gradually detached and rotated to an upright angle by an entirely natural process of weathering. Other outcrops where this has happened are visible

A panel of carvings at Holly Canyon near Hovenweep is transformed by remarkable lighting effects, here photographed by the author on the morning of their discovery by Ray Williamson at the June solstice in 1979. A "sun dagger" pierces a carving of three concentric rings, a symbol still widely recognized by the Pueblo today as standing for the sun.

throughout Chaco Canyon. This explanation would mean that the Anasazi had simply observed the unusual light forms behind the slabs and recorded them by carving the spiral patterns, without taking any steps to create an observatory by moving or modifying the stones.

Even if we prefer the simpler explanation, the Fajada Butte "sun dagger" is still a vital discovery, since it demonstrates that the Anasazi were fully aware of the changing motion of the sun through the seasons. Perhaps the site high up on the Butte was a solitary place of retreat for a sun priest, similar to the officials who performed solstice observations at Hopi and Zuni in recent times.

Were the problems of existence in Chaco Canyon therefore comparable to those of the Hopi farmer, for whom the sun's seasonal movements were of such practical use in the timing of agriculture? If we imagine that the inhabitants of Chaco really were like the Hopi and Zuni in their outlook and way of life, how do we account for the phenomenal energies that led to the construction of the great towns in the eleventh and early twelfth centuries? What can have possessed the canyon dwellers to reach such heights of engineering skill and communal laboring?

THE CHACO ACHIEVEMENT

The Chaco planners worked on a scale that bears no comparison with modern pueblos. No standard units of length are prescribed for the Hopi community as a whole (though the "pace" was apparently common in the construction of individual homes a century ago). Buildings often face the four cardinal directions, but none of the modern pueblos exhibit an overall unified plan like that of Pueblo Bonito and the other Chaco towns.

Domestic dwellings are usually raised by a small group of relatives, with occasional cooperation from clansmen and neighbors; larger work teams are only recruited for raising special public structures, such as *kivas*. In contrast, even the smallest of the Chaco towns, known as Wijiji, involved an extraordinary degree of forethought and manpower. Wijiji was built in one operation and is thought to incorporate twenty-five hundred trees in its structure. The walls of one of the largest towns, Chetro Ketl, probably contain well over fifty million pieces of building stone. To support the mighty wooden columns of the *kiva* at Chetro Ketl there were great stone discs, said to weigh anything from a thousand to fifteen hundred pounds each. They were evidently dragged into the canyon from an outside source. Human energies were harnessed here at a level unmatched in the long history of the Southwest.

A striking aspect of this activity in the eleventh-century Chaco towns was the provision of "public works," including networks of roads and irrigation systems. The planning and maintenance of these facilities were presumably shared among the thirteen towns, all of which are huddled close to the north wall of the canyon. It was this cliff which most effectively captured the runoff from violent summer thunderstorms. At the entrances of clefts and gullies where the water gushed down, floodwater farming similar in principal to the techniques of the Hopi farmer must have been possible. At Chaco, though, the building of water-control features such as check dams, irrigation ditches, and bordered plots was pursued on a scale unheard of among the Hopi in recent times. The ditches ranged from specimens a few feet wide and deep to genuine canals, lined with masonry, up to fifty feet across.

Once the rainstorms broke, the water diverted from the clifftops and the Chaco River was fed partly into reservoirs for everyday use, and partly into earth-bordered gardens located on the canyon floor. Only one such field near Chetro Ketl has actually been investigated, and because it dates to a late period, it may not be typical of Chaco agriculture. Nevertheless, the field is remarkable for the care with which it was prepared, including an almost perfect leveling operation and a layout of earthen banks at

Distant view of Pueblo Bonito from Casa Rinconada on the opposite side of Chaco Canyon.

right angles scarcely deviating from ninety degrees. The orientation of the field system to within a few degrees of the summer solstice sunset may be a coincidence, but the hint that simple geometry and astronomy influenced the planning of fields as well as towns is intriguing.

Could six thousand people have been supported by ingenious irrigation systems of this kind? Despite the intensive methods of crop raising, recent archaeological surveys indicate that less than two percent of the total area in the canyon suitable for flood farming was actually exploited at any one time. According to one estimate, this tiny proportion of the available land would have sustained eighty-two people! The elaborately constructed gardens may have been an important "cushion" in times of shortage or stress, but were inadequate for the everyday needs of subsistence. Much of the food must have come from elsewhere.

The existence of ancient roadways was noticed as long ago as 1900, but the full scale of the Chaco network was only realized after detailed aerial mapping of the canyon was accomplished during the 1970s. The use of a variety of photographic techniques, including infrared film and electronic image enhancement, revealed segments of roads from the air that were barely visible at ground level.

It is accurate to describe them as roads, for originally they were set out in remarkably straight lines. Their surfaces were cleared of rubble, and in some places the edges were lined with curbs of masonry. More than two hundred miles of roads have been plotted so far. This network connected the major Chaco towns with each other and also led to satellite settlements located as far as sixty miles away from the canyon. The "satellites" were generally located near land with good agricultural potential; their architecture was similar to the classic building style of the great towns, and included large *kivas*. These communities and the roads that connected them may indicate the way in which the authorities in the canyon coordinated the agricultural production of the entire region. Food exchanges may have taken place in the course of ceremonies staged inside the huge *kivas*, or perhaps even in markets held in the open plazas at Pueblo Bonito and other towns.

The crucial question is whether a class-ridden society controlled by powerful chiefs and priests was essential for the working of the Chaco system and the growth of sunwatching practices. The opinions of researchers are sharply divided on this issue, yet so far the search for signs that individuals of high status were in charge of the Chaco towns has proved disappointing.

No cemetery areas specially set aside for the dead have ever been recognized in the canyon, and excavations at the town sites have yielded only about 325 burials. While these graves may represent a privileged minority, most of the burials seem humble and undistinguished in material terms. The only exception was a group of a dozen or so individuals buried together under the floors of four adjoining rooms in the old north sector of Pueblo Bonito. They were buried with hundreds of turquoise beads, wooden canes, conch-shell trumpets, and a painted flute. Some of this material could represent ritual or shamanistic aids rather than the regalia of a chief.

In any case, there were certainly no great palaces or huge bulk-storage facilities suitable for the pomp and power of authoritarian leaders. Was the running of the towns and of the entire Chaco network managed by community organizations, operating mainly by collective, not hierarchical, decisions?

It is even difficult to distinguish the character of the town dwellers from those of the hundreds of small eleventh-century villages scattered throughout the length of the canyon. The villagers ate and drank from the same kind of vessels as the town people, although certain exotic painted wares and other luxury items were found only in the towns. Fragments of painted wood likely to have formed parts of ceremonial masks or festishes were recovered from town and village alike. The location of Casa Rinconada on the south side of the canyon, away from any major settlement but close to several villages, suggests that the same ritual acts may have united town and village people.

Indeed, if we suppose that great disparities in wealth and a harshly coercive leadership did not exist, then shared sacred values may have been crucial for drawing together the efforts of the scattered communities. The origin of sunwatching practices was probably closely involved with the development of these ceremonial activities. No doubt a widely observed calendar would have helped to solidify the outlook and customs of the dispersed farming population. A solar calendar may also have developed in response to the unpredictable nature of agriculture in some of the outlying, marginal areas of the far-flung Chaco network.

The individuals who watched for the solstices from the solitary shrines on the summit of Fajada Butte or on the cliff top at Penasco Blanco must have been influential in the community, as

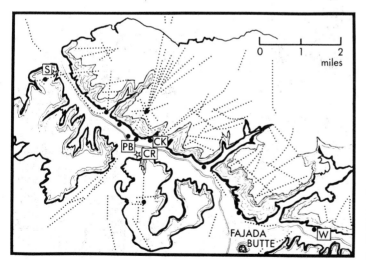

The major "towns" and the ancient road system of Chaco Canyon. PB is for Pueblo Bonito, CK is for Chetro Ketl, CR is for Casa Rinconada, W is for Wijiji, S is for the "supernova" painting.

were the Zuni *Pekwin* and the Hopi Sun Priest in recent times. But there is no evidence that these officials necessarily belonged to a special cult or an astronomer-priesthood separated from the rest of society.

To put it simply, the energies concentrated in the great towns of Chaco may indicate that their populations were organized in more complicated ways than those characteristic of the modern Pueblo. Nevertheless, an oppressive, rigid hierarchy similar to that of Californian tribes such as the Chumash seems unlikely. Both recently and in prehistory, the growth of astronomical practices was probably intertwined with the subsistence problems of the ordinary farmer. The sacred ceremonies were performed for his benefit, and did not "line the pockets" of a priestly class.

THE FATE OF THE "OLD ONES"

This reconstruction of life in Chaco Canyon less than a thousand years ago is obviously still tentative, since the details of a particular social order are always difficult to establish from archaeological evidence alone. One certain fact is that the system of the Chaco people did not last for long. They undertook their most ambitious architectural projects in the early eleven hundreds. Only about a century later it was all over: the classic Chaco style of building ceased.

Many writers have envisaged the end of the Chaco people in terms of a catastrophe. Their disappearance from the canyon is usually attributed to the effects of a drought documented in

The steeply eroded banks of the Chaco River make it useless for irrigation today.

several regions of the Southwest during the mid-twelfth century. Their situation may have been worsened by wasteful farming and building practices. (For example, estimates of the total number of trees incorporated as floor and wall beams in all thirteen towns vary from one hundred thousand to one million; the removal of pine and juniper cover from the broad mesas may have hastened the process of soil erosion). This vision of environmental disaster has an understandable fascination for us today.

However, the latest archaeological evidence suggests that occupation of the canyon actually continued without serious interruption until well into the thirteenth century. Many of the changes previously thought to reflect a drastic alteration in the canyon dwellers' way of life can now be explained in less dramatic terms. For instance, there seems to have been a slow shift in the organization of trade and subsistence toward Mesa Verde and the Four Corners region. The Chaco people were not at the mercy of environmental changes; they seem to have faced a slow decline in economic importance, rather than a sudden crisis.

In any case, it is probably a misconception to think of the Chaco people as a "vanished race." The precise origin of the Hopi and Zuni people is a complex topic still debated by anthropologists; however, the evidence of solar observations among the Anasazi does suggest a number of striking parallels with later beliefs and practices.

In the Four Corners area, a few examples of stone towers with astronomically aligned wall slots seem to anticipate the anthropologists' descriptions of similar structures at Zuni and Cochiti. The grand scale of architectural planning at the major Chaco towns was unique, but the orderly way in which features were arranged in and around *kivas* and plazas invites comparison with the modern pueblos. Moreover, the discovery of the remarkable "sun dagger" device atop Fajada Butte hints at the existence of officials similar to the Pueblo sunwatchers with their special awareness of the sun's movements. The other solar observing sites so far identified in the canyon are not nearly as convincing; nevertheless, the appearance of rock paintings still widely recognized as astronomical symbols by modern Pueblo people strengthens the case for these sites.

Whatever controversy is stirred by this exciting new research, the weight of the evidence suggests that elements of Anasazi skywatching lore did persist, and that a thread of continuity connects over a thousand years of sacred traditions in the Southwest. It is also clear that an oppressive hierarchy of priests and chiefs was not essential to give rise to intricate sky beliefs and practices.

Night falls over the pyramid known as the House of the Magician at the Mayan site of Uxmal, Yucatán, Mexico, dating to about 800 A.D.

PART IV

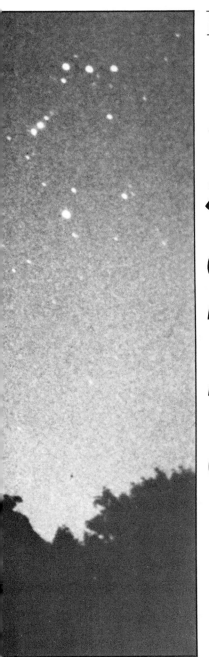

Skywatchers of the Tropics:

The Ancient and Modern Maya

12. *Under the Tropical Sun*

In the tropics, the ancient skywatchers of the New World reached their most advanced level of skills. The previous chapters have indicated the richness and variety of cultural traditions in pre-Columbian North America, yet the accomplishments of the tropical civilizations were certainly of a higher order.

These achievements were the product of many centuries of slow growth. During the same era as the rise of Imperial Rome, the scattered farming communities of central Mexico were gradually transformed into the first bureaucratic state in the Americas, which boasted far-ranging trade connections and impressive public monuments of stone. As Charlemagne was crowned emperor of the West in Rome in 800 A.D., the intellectual development of Central America was at its peak under the Mayan civilization. By that time, the population of the Mayan lowlands was gathered by the thousands in great settlements which had many of the essential features of a city. There the priesthood must have been a formally organized institution, and one of its concerns was writing and keeping calendrical and astronomical records.

The invention of numbers and hieroglyphs was a remarkable advance, clearly distinguishing the tropical cultures from other Native Americans. Preliterate societies such as the Chumash and the Zuni, with their notched sticks and tally cords sometimes referring back to centuries of accumulated traditions, were obviously capable of sophisticated astronomical concepts and observations. Yet when the ancient Mexicans learned to record dates within a chronological framework (the so-called "Long Count" system, which seems to have been devised during the first century B.C.), a new dimension was added to their skills. The Maya developed number techniques similar in principle to those of the Babylonians. By consulting written records and counting the intervals between sky events, the Maya were eventually able to anticipate the timing of lunar eclipses and planetary movements accurately. These new powers, along with the knowledge of reading and writing, meant that skywatching was increasingly confined to an elite group who interpreted the records and ensured their transmission from one generation to the next.

Why did this unique development of ancient American writing and mathematics take place in the tropics? To understand the background of this achievement, we first need to appreciate the special conditions of skywatching in the tropical zone. An observant traveler to this region will notice how different the sky is here than in the north. Indeed, certain traditions of skywatching could only have developed near the equator because of the special nature of celestial movements here.

Perhaps the best example is the use of stars for accurate, long-

THE ROOF OF VOYAGING

Opposite, *in the tropics, the sun shines vertically overhead on two days of the year, creating an impressive noon spectacle.*

distance navigation, a skill which it was impossible to practice outside the tropical zone. At the time of European contact, the Pacific navigators of Polynesia were said to be engaged in planned voyages of up to five hundred miles in length. The surprising accuracy of their methods was soon noted by explorers such as Andia y Varela, a Spaniard who visited Tahiti in 1774–1775.

> When the night is a clear one they steer by the stars . . . not only do they note by them the bearings on which the several islands with which they are in touch lie, but also the harbors in them, so that they make straight for the entrance . . . and they hit it off with as much precision as the most expert navigator of civilized nations could achieve.[51]

Polynesian navigators rarely used charts or recording devices. But until 1900, Marshall Islanders would consult "stick compasses" before setting out. These memory-jogging aids symbolized visible ocean currents and swells encountered between islands.

The secret of the Polynesian navigators was their ability to memorize a long sequence of rising or setting stars which would be seen, one after another, above the same spot on the horizon. The chain of stars represented a kind of compass that must have been developed over many years of experience and experiment. It was, of course, essential to know beforehand exactly where one's destination lay and which stars were associated with it. The navigators were also guided by their expert understanding of tides, winds, wave patterns, bird behavior, and even ocean smells; according to the description of a navigator from the Gilbert Islands of Micronesia, "a compass was another of the things he could do without. He had his guiding stars for every night of the year and every state of wind or current, he said, and these, with his sense of smell, made him independent of white men's inventions."[52]

Basic sky movements at the equator, left, and in the northern zone, right. In the tropics, the horizon and the zenith overhead are the fundamental reference points because of the vertical motion of objects in the sky. Farther north, the universe appears more complex, because it rotates around the Pole Star at an inclined angle to the horizon.

Another navigation technique required a knowledge of which stars "ruled over" individual islands—in other words, which stars would appear in the zenith directly above an observer stationed on a particular island. If the voyager could recall this information, it was simply necessary to steer north or south until the same "ruling star" passed overhead; then the navigator would be on the same latitude as the island, and could set course either due west or east toward it. No doubt this method was useful for

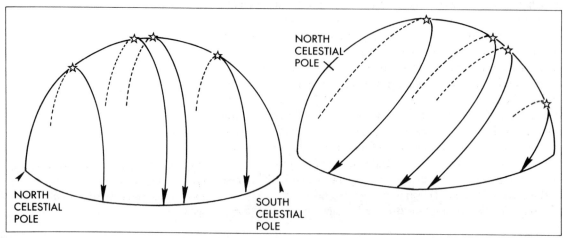

NORTH
CELESTIAL
POLE

NORTH
CELESTIAL
POLE

SOUTH
CELESTIAL
POLE

finding a way home after a long-distance voyage of exploration.

The key reference points of the Polynesian navigators were either "ahead" to the horizon or "up" to the zenith above. Their methods were feasible because of the simple character of sky motions at the equator. The rotation of the earth is mirrored by the movements of the sun and stars; if the earth's geographical lines of latitude were projected onto the sky, the stars would be seen to move exactly along them. A star rising due east passes through the zenith overhead, and sets due west. The paths of the rising or setting stars are all vertical in relation to the horizon; it was this "straight up" or "straight down" motion that made the Polynesian star compass practical.

In the northern latitudes, on the other hand, the whole celestial framework is tilted at an angle from the observer as it pivots around the Pole Star. The motions of the stars are not simply "up" or "down," but also "around," following oblique paths through the sky. This is critical for the northern navigator; if the direction of a single star is followed for any length of time, it will quickly throw him off course because of its horizontal as well as vertical movement. For instance, a sailor crossing the North Atlantic will deviate a whole ten degrees from his intended bearing in the course of only a single hour. This explains why accurate celestial navigation without modern instruments is only possible in the tropics.

The perfect symmetry of the sky at the equator made it easier for the navigators of the Gilbert Islands to identify and memorize large numbers of stars. They visualized the night sky as a huge canopy overhead, and called it *uma mi borau*, which means literally "roof of voyaging." Like an ordinary house roof, it was divided by imaginary rafters and cross beams. The positions of the rafters corresponded to the paths of prominent stars such as Rigel, Antares, and the Pleiades group.

Canoes of the Gilbert Islands in Micronesia. Larger versions of this basic design were built for ocean voyages.

An astronomical official of the Kenyah in Borneo measures the length of the sun's shadow with the aid of a gnomon. It was critical for the Kenyah to forecast the arrival of the dry season, when forest cover had to be burned to make way for new rice fields. The official had no clear idea of a solar or lunar calendar, but simply knew the different appearances of the noon shadows.

By thinking of the sky as an enormous symmetrical framework, the Gilbert navigator provided himself with a simple celestial reference system. He could locate a star anywhere in the sky by its relationship to the fanciful grid of beams and rafters. Indeed, the instruction of the novice, which could last as long as seven years, began not under the night sky but under the roof of the village meeting house. There he had to commit to memory dozens of different stars, constellations, and nebulae, reciting their relative positions among the roof beams at any particular season of the year. It must have been a formidable and solitary task, for the star secrets were jealously guarded by the elite navigators; indeed, each instructor devised his own private constellation groupings.

The Polynesian navigators were not the only tropical peoples to adopt the framework of a house as an appropriate model of the sky. Many native South American tribes living close to the equator arrange their temples and the entire ground plan of their villages as a reflection of the orderly pattern overhead.

A striking example of this practice is provided by the Kogi, who live in the foothills of the Colombian Andes, and have been studied for many years by the anthropologist Gerardo Reichel-Dolmatoff, who teaches at the University of California. The Kogi temple is oriented with its doors facing east–west, the major direction of celestial movement. The Kogi *mama*, or shaman, measures the proportions of the circular ground plan using a cord attached to a central stake. The outer ring of wall posts is planned so that four prominent uprights mark the solstices. The locations of the two midsummer posts are established by direct observation of the sun, and two midwinter posts are then measured off at corresponding points on the opposite side of the ring. Finally, the positions of the solstice posts are somehow adjusted so that they are also believed to represent the intercardinal directions. In this way, the ground plan of the temple becomes a perfectly symmetrical diagram of the universe.

The superstructure of the temple reflects other aspects of Kogi cosmology. The four tiers of roof beams represent the four layers of the heavens, and the framework of the temple is also imagined to continue underground in an exact mirror image. In Kogi eyes, then, the temple—like the world itself—resembles a pair of cones joined together at the base; it is at the junction of the cones at the fifth, or middle, level where human existence takes place. The appearance of the imaginary double cones reminds the Kogi of the spindles they use for weaving, and so the erection of the first spindle (or zenith post) heralds the creation of the human world:

> In the beginning of time, the Mother Goddess took a spindle and pushed it upright into the newly created and still soft earth, right in the center of the snowpeaks of the Sierra Nevada, saying "This is the central post. . . !" And then, picking from the top of the spindle a length of yarn, she drew with it a circle around the spindle-whorl and said: "This shall be the land of my children."[53]

The Kogi shaman thus reenacts the moment of creation when he sets out the temple with the aid of a cord fixed at the center stake.

Furthermore, the sacred "weaving of the sun" is also dramatized at each of the solstices. On these occasions, the shaman removes a potsherd covering the small hole at the apex of the conical temple roof. A spot of sunlight shines on the floor below and, from dawn to sunset, crosses the space between two pairs of carefully positioned fireplaces. The sun is said to be "weaving a fabric on the loom of the temple."

Many South American peoples share the Kogi preoccupation with the mirrorlike patterning of space around a central, vertical axis. In fact, most native villages of the central Brazilian plateau are arranged around an east–west line dividing the population into two halves, or moieties, similar in principle to those encountered in the pueblos. The famous French anthropologist Claude Lévi-Strauss has vividly described the way of life in these Brazilian communities. Writing of the Bororo, for instance, he compares the activity of their moiety groups to a "ballet," in which the two divisions "strive to live and breathe through and for the other; exchanging women, possessions and services in fervent reciprocity; intermarrying their children, burying each other's dead, each providing the other with a guarantee that life is eternal, the world full of help and society just. . . ."[54]

Among the Apinayé of southeastern Brazil, the balance of authority between the two village moieties is dramatized in an elaborate ritual ball game held about every ten years at the time of the September equinox. The two teams, each drawn from a different moiety, line up on either side of the east–west pathway bisecting the village. Exactly as the sun begins to rise along the pathway, a large rubber ball is passed from the north to the south line of players. While it cannot have been very entertaining as sport, the "ball game" solemnly justifies the division of the tribe into two halves. Just as the equinox sunrise divides up the year between the solstices, so this ceremonial passing of the ball expresses the balance between the two sections of the community.

In the tropical latitudes, it is common to identify human affairs closely with movements in the sky, as the Apinayé ball game illustrates. The symmetry of the heavens along an east–west axis deeply influences the social and religious thinking of peoples situated near the equator. The vertical movement of the sun and stars encourages the notion that the observer and his or her community are located at the center of everything, where all natural forces and conflicts are precisely balanced. An emphasis on the "up and down" and "here and now" in the organization of space and religious beliefs is understandable.

Farther north, it is impossible to cling to such a comfortable identification with the sky. As the celestial paths shift from "up and down" to "up and around," the significance of the human

A *Tukano of northwest Brazil in ceremonial costume, photographed between 1903 and 1905.*

observer in the cosmic scheme changes. The asymmetrical motion of the sun and stars around the pole suggests a source of supernatural power located *outside* the observer and his or her immediate surroundings. Explanations and images of the universe become more complicated than the spindle-whorl of the Kogi. The link between human affairs and divine forces is less direct and secure.

SUN LINES IN ANCIENT PERU

In the tropics, the vertical axis connecting an observer to the zenith overhead is vital as the link between the earth and other worlds. This connection can be established anywhere, not just at the center of a temple or a community, by erecting a sacred pole to communicate with the supernatural beings above. For example, when the Sherente tribe of eastern Brazil experience prolonged drought, their elders decide to raise a ten-meter-high pole which they call the "road to the sky." This pole is erected at night in a forest clearing away from the village, so that representatives of the two moieties can climb to the top and converse with the spirits (who are identified with prominent stars). Besides weather information, the Sherente believe they can find out how long they are to live by questioning the souls of their dead relatives at the top of the pole.

These examples suggest that the astronomers of the ancient tropical civilizations would pay particular attention to the "up," or zenith, direction. They would notice the day on which the sun shines vertically overhead because of the dramatic moment at noon when all shadows temporarily vanish. Upright markers (gnomons) or other shadow-casting devices would aid the observer in anticipating the approach of this day.

Such markers were once common throughout the Inca empire of Peru, according to the chronicle compiled around 1600 by Garcilaso de la Vega, son of a Spanish captain and the nephew of Huayna Capac, the eleventh ruler of the Inca. In Garcilaso's *Royal Commentaries of the Inca,* he describes a particularly sacred group of gnomons located near Quito, very close to the equator. At the equinoxes in March and September, the sun would rise due east and set due west, passing through the zenith at noon; or, as Garcilaso puts it quaintly in the *Commentaries,* the markers "afforded the sun the seat he liked best, since there he sat straight up and elsewhere on one side."[55] At other locations either to the north or south of the equator, the two days of the zenith passage did not coincide with the equinoxes. To predict the arrival of these dates, Garcilaso states that the Inca watched their markers until "at midday the sun bathed all sides of the column and cast no shadows at all. . . . Then they decked the columns with all the flowers and aromatic herbs they could find and placed the throne of the sun on it, saying that on that day the Sun was seated on the column in all his full light."[55]

The Church of Santo Domingo, Cuzco, is built on top of the original Inca walls of the Coricancha, the Temple of the Sun.

If the ancient observers had traveled far beyond the limits of the Inca empire and reached either the Tropic of Cancer or the Tropic of Capricorn, they would have found that the sun shone overhead on only one day of the year, at midsummer and midwinter respectively. Beyond the tropics, of course, the phenomenon of the "shadowless day" never occurs.

From the accounts compiled by Garcilaso and his contemporaries, we know that the Inca developed other astronomical skills besides the observation of noon shadows. The layout of the ancient capital of Cuzco, located in the high Andes 13° south of the equator, was extensively replanned by the Inca ruler Pachacuti in the mid-sixteenth century. Somewhat like a Brazilian village, the city was bisected into a northern and southern moiety. There was a further subdivision into quarters, symbolizing the four directions of the empire, while the four thousand or so houses of the city were neatly arranged in rectangular wards separated by narrow streets. At the heart of the city was the Coricancha, the Sun Temple, where several magnificent gold images of the sun were installed in its innermost shrine. Other rooms were dedicated to the worship of the moon, the stars, the thunder, and the rainbow.

From the Coricancha, a system of forty-one imaginary lines, or *ceques*, radiated outwards through the city and surrounding countryside. The path of each *ceque* was marked by holy shrines or *huacas* located at intervals along its length. The *huacas* were a diverse collection of over three hundred venerated sites including springs, fountains, bridges, houses, hills, caves, and legendary tombs and battlefields. The responsibility for their upkeep was assigned to particular family groups within the capital. The whole *ceque* system was the basis for organizing many pious and practical activities, including irrigation and the prediction of winds. Despite much confusion in the chroniclers' accounts, it is clear that some *ceques* also served as astronomical sight lines. They were aimed toward distant towers or pillars, positioned as horizon markers on the crests of the hills surrounding Cuzco.

The Incas celebrate the solstices; two illustrations dating to about 1600 from the native chronicle of Felipé Huamán Poma de Ayala, an Inca of noble descent.

Huacas, or sacred shrines, often marked by crosses or stones, abounded in the countryside surrounding the Inca capital, Cuzco.

The structures on the hilltops were all demolished by the Spanish invaders, but in many cases the ancient place-names assigned to the *huacas* have survived. By plotting the present-day locations of these spots, it is possible to reconstruct the course taken by a number of astronomically significant *ceque* lines. During the late 1970s, for instance, Anthony Aveni of Colgate University and Tom Zuidema of the University of Illinois demonstrated that a pair of pillars on the southwest horizon had once marked the December solstice sunset when viewed from the center of the *ceque* system, the Coricancha temple.

The testimony of the early Spanish chroniclers indicates that an even more important date in the ancient calendar was associated with the beginning of the planting season in the valley of Cuzco. This was in mid-August, a time that native Andean peoples still think of as the "opening up" of *Pachamama*, the Earth Mother underfoot, who is impregnated by the farmer's seeds. In the Inca mind, this belief was apparently linked with the date in mid-August when the sun passes vertically *underneath* the feet of the observer at noon. This subterranean reference point exactly opposite the zenith is usually referred to as the "antizenith." So the sun's crossing of the antizenith can be thought of as the reverse of the zenith passage phenomenon, which occurs six months earlier.

Of course, no one could observe the "antizenith passage" of the sun directly. Its timing could be established quite easily, however, by projecting the point of sunrise on the day of zenith passage over to the horizon diametrically opposite. This spot would then mark the sunset point on the antizenith day.

The existence of this double alignment to the zenith sunrise–antizenith sunset was, in fact, convincingly demonstrated by Aveni and Zuidema. Their research shows that the "up and down" symbolism of the axis connecting the observer to the zenith above and the antizenith below played a key part in the agricultural and ceremonial schedule of the Inca, and in the whole elaborate *ceque* system itself.

Now that investigators realize the importance of the vertical direction of space in native Andean thought, fresh efforts are under way to explore one of the great mysteries of the New World, the famous markings on the plain of Nazca. These markings are located near the Peruvian coastline, about 250 miles southeast of Lima, in one of the world's most arid landscapes. The level pavement of pebbles and boulders was formed by centuries of erosion from the Andean foothills to the east.

On this surface, the people of the Nazca culture (dating from about 300 B.C. to 700 A.D.) created a vast network of geometrical patterns by a simple technique. The lines were made by removing a top layer of pebbles, which had been oxidized to a dark red-brown by unceasing exposure to the sun. When this layer was removed, the lighter subsoil underneath formed a contrast with the surrounding pavement. But the contrast is so subtle and the

Opposite, *possible astronomical structures at Machu Picchu, the "lost city" of the Incas, currently under investigation by David Dearborn and Raymond White of the University of Arizona. The legendary Intihuatana Stone, above, though apparently not an observing instrument, may have marked the spot where Inca astronomers stood to watch the horizon. Two windows in the Torreon building, below, seem to have been positioned for observation of the June solstice and also of constellations around the time of zenith passage.*

Many of the Nazca lines are almost invisible at ground level. From a hilltop vantage point the remarkable straightness of some lines is strikingly apparent.

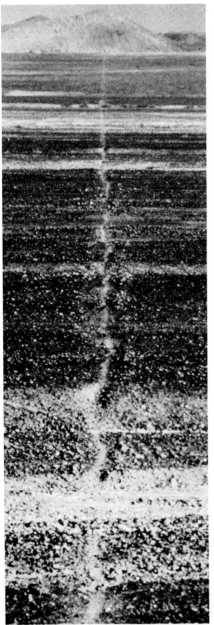

This giant bird design on the Nazca desert is crossed by a straight line, parallel to the outstretched wings, running in the direction of the solstices. But less distinct lines cross the bird in other directions, too.

scale of the markings so huge that their existence was only discovered when scheduled airline flights began to cross the Peruvian highlands during the 1930s.

A decade later, Paul Kosok, an archaeologist from Long Island University, heard rumors of the strange sights seen from the air, and began investigating the mystery. In 1946 Kosok was joined by Maria Reiche, a German-born mathematician who has devoted her life to studying, surveying, and protecting the Nazca lines. Now in her late seventies, Reiche continues her solitary task of mapping the desert designs before they are obliterated by the casual footsteps and car tracks of tourists.

The pottery fragments scattered across the desert suggest that the markings were probably made by people of the local Nazca culture at least a thousand years before the Inca created the *ceque* system of Cuzco. However, since many of the Nazca lines radiate from central points, their resemblance to the *ceques* may offer a clue to their purpose.

The possibility that some of the lines are astronomically significant, as in the case of the *ceques*, was first considered many years ago. After Kosok observed the June solstice sun setting along one of the lines, he became convinced that the Nazca drawings represented "the largest astronomy book in the world." His idea that the designs formed a kind of gigantic calendar has been supported consistently by Reiche, who discovered that a number of huge animal designs etched on the desert are apparently aligned to the solstices.

Nevertheless, the only systematic study of the problem yet published offers the surprising conclusion that the lines were probably *not* astronomical. In 1968, Gerald S. Hawkins, fresh from his triumphant "decoding" of Stonehenge, organized an accurate survey of a small section of the Nazca drawings, using aerial photogrammetric techniques. From the resulting plans, Hawkins measured the angles of 186 directions defined by the desert drawings. He then "fed" these measurements to the same computer program he had developed for Stonehenge. The analysis showed that only 16 of the 186 directions were likely to be

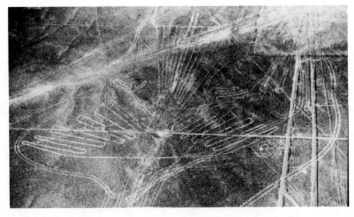

deliberate alignments, leaving more than 80 percent of the markings unexplained. A few of the 16 did look promising—such as the solstice line that passed right through the drawing of a giant bird—yet in most cases there was nothing to distinguish the possible astronomical lines from others that pointed nowhere near the sun, moon, or stars. Hawkins therefore concluded that the small number of significant lines was probably due to chance. The last words of his report, published in 1974, are that the markings "remain an unsolved mystery in the desert."[56]

However, this was not the end of Hawkins' involvement with Nazca, for he also collaborated with zoologist and filmmaker Tony Morrison, who has traveled widely in the Andes during the past twenty years. The result of their efforts was a dramatically illustrated book called *Pathways to the Gods*, published in 1978. In this book Morrison describes his contacts with highland Peruvian peoples such as the Aymara, some of whom still make ritual processions along straight-line pathways resembling the *ceques* and *huacas* of Inca tradition. A few, such as those associated with the huge animal drawings, might be astronomical, but Morrison follows Hawkins in believing that the majority are not.

At the time of Hawkins' earlier work, the significance in Andean lore of a number of celestial phenomena—notably, the zenith and antizenith passages of the sun—was not widely known. Thus, Hawkins' pioneering analysis proved only that the Nazca lines did not indicate the same horizon events which he believes were observed at Stonehenge, such as the solstices and lunar extremes.

With their research on the Cuzco *ceque* lines and other studies of contemporary Andean folklore behind them, Aveni and his colleagues are currently mounting new efforts to investigate the meaning of the Nazca lines. As this book goes to press, they are surveying the desert for evidence of the variety of purposes which the lines probably served, as routeways, sacred paths, or perhaps even as a form of artistic expression. They are also examining the markings for alignments to events known to be of interest to

Air view of a small section of the Nazca lines.

ancient skywatchers in the tropics, such as the rising and setting of the sun on the days of zenith passage. If, indeed, an astronomical interpretation of some of the lines *is* eventually accepted, then the Nazca case will emphasize the striking differences between prehistoric observing traditions in the northern latitudes and in the tropics.

CONQUEST AND THE CALENDAR

The opportunities for future surveys and discoveries in South America are obviously vast. By comparison, scholarly knowledge of the part played by astronomy in the civilizations of Central America is more complete and better established. The importance of the zenith, for example, has been recognized for a long time—in fact, ever since the Spaniards encountered the Aztec of Mexico, who timed the ending of their most crucial calendar cycle by the midnight passage of the Pleiades overhead. In 1577 King Philip II of Spain issued a decree instructing the authorities of each city in the Indies to report their latitude and the two dates of solar zenith passage. The purpose of the decree was to enable the Spanish to control any outburst of native religious fervor connected with sun worship.

In the southern part of Central America, the first zenith passage day marked the traditional beginning of the year in many regions, and also coincided with both the start of the rainy season and the time to begin planting. During the last century, the approach of this day was anticipated with an accuracy that surprised European observers. Even today, the inhabitants of remote villages in highland Guatemala still celebrate the two zenith passage dates with elaborate parades and rituals as a new phase of the agricultural cycle begins.

The zenith passage of the sun creates a dramatic moment at the palace of Mitla in Oaxaca, Mexico, probably dating to the 13th–14th centuries A.D. The vertical noon sun precisely outlines the courtyard, while the façades are completely in shadow.

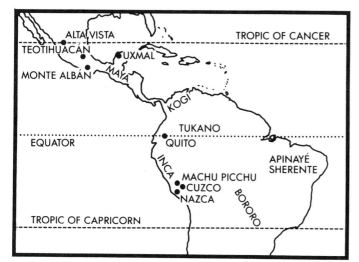

Map of the tropical sites and peoples discussed in this chapter.

Shadow-casting devices such as gnomons were not the ony way of determining the arrival of the shadowless day. At two ancient sites in Mexico, the remains of carefully constructed, narrow vertical tubes open to the sky have been identified. If an observer were positioned inside one of these structures near the bottom of the tube, the advent of the sun vertically overhead could be watched at noon over the course of several days or weeks. The observer could either look straight up at the patch of sky defined by the mouth of the tube, or else wait for a spot of sunlight to shine down the tube and penetrate the darkness at his feet.

The most interesting of these "zenith tube" observatories is located in one of the great centers of Central American civilization, the Valley of Oaxaca in the southern highlands of Mexico. Oaxaca today is a prosperous manufacturing and commercial city of more than two hundred thousand inhabitants, lying at the junction of three major river tributaries and surrounded by steep-sided, forested mountains. Some parts of the valley receive so little rainfall that the years of failed harvests outnumber the successful ones.

Despite these conditions of almost permanent drought, the valley was one of the earliest and liveliest centers of culture in the New World. Farming began here at an early date, with tentative experiments at raising pumpkins and a primitive variety of maize reaching back as far as 7000 B.C. It was not until about 1500 B.C., however, that a settled existence dependent on agriculture finally developed. The population of the valley was spread among a dozen small villages, and expanded gradually as the new way of life took root.

The smooth development of village life in the Oaxaca Valley was abruptly transformed around 600 B.C., when the inhabitants organized the construction of a huge complex of ceremonial buildings on top of a mountain known today as Monte Albán. Although there was no water available, and a climb of nearly fifteen hundred feet from the valley floor was necessary, the

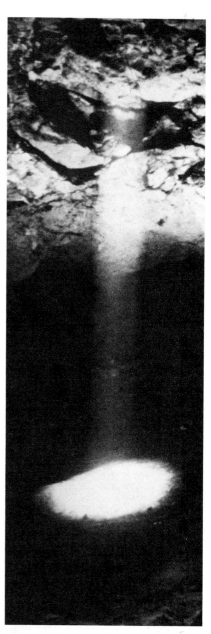

The arrival of the sun overhead was so important that ancient Mexicans built vertical "zenith tubes" to observe the phenomenon. Here, sunlight penetrates to the bottom of a suspected zenith tube at the ruins of Xochicalco.

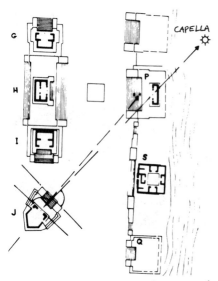

Survey by Horst Hartung shows part of the Monte Albán plaza, with the alignment from Building J to Building P.

Below, *the plaza at Monte Albán, with the oddly shaped Building J in the foreground.*

builders could not have chosen a more dramatic or commanding place for the foundation of a city. At the height of Monte Albán's power and prosperity, around 300–600 A.D., it is estimated that between thirty thousand and seventy thousand people were settled on the slopes of the mountain and on the neighboring hills. Their dwellings surrounded a huge ceremonial plaza about one thousand feet long and seven hundred feet wide, flanked by the principal temples and residences of the city. The effect of this great open space, elevated so high above the landscape that one feels close to the clouds, is still overwhelmingly impressive today.

All the buildings in and around the plaza share the same general orientation slightly to the east of north, with one glaring exception. Near the plaza's southern end stands one of the most oddly shaped structures of the ancient world, known to archaeologists by the unimaginative name of Building J. No two sides or angles of Building J are equal, and its awkward configuration resembles the stranded prow of a battleship. Besides its peculiar plan, Building J is also exceptional because it is positioned at an angle about 45° out of line with the rest of the buildings at Monte Albán. Could the existence of astronomical alignments explain some of its strange features?

The front stairway of Building J is so constructed that a sight line perpendicular to the steps passes across the plaza directly through the front doorway of another edifice, Building P, to the northeast. If projected to the horizon, this line also corresponds to the point where the bright star Capella was seen in the dawn

sky for the first time each year. This day also happened to coincide with the first zenith passage of the sun, so the earliest annual appearance of Capella could have served as a kind of warning signal, telling the priests that the sun would arrive overhead later that day. In other words, a concern with the zenith passage event may have inspired the architects' odd positioning of Building J.

This argument is greatly strengthened by the fact that they apparently arranged for a "zenith tube" to be built into the steps of Building P. A narrow gap in the stairway allows the visitor to scramble into a gloomy chamber big enough to accommodate one or two people. At the end of the chamber is a narrow recess situated underneath a square shaft, about six inches across, that extends vertically through the stairway above. The sun is visible through this perpendicular shaft for several weeks around the time of the zenith passage, but within a day or two of the actual date, the sun shines down into the chamber for a noticeably longer period of time. The combination of the "warning star" Capella and the solar image in the zenith tube must have enabled the Monte Albán observers to fix the date of the first zenith passage with considerable accuracy every year.

The ancient skywatchers of Oaxaca were not disinterested scientists pursuing observations for their own sake, however. Instead, they belonged to an aggressive, vigorously expanding empire with far-reaching commercial and military connections. From its earliest beginnings, Central American astronomical and calendrical knowledge was always interwoven with political and religious values.

Building J presents a good case in point. One of the many remarkable features of this structure is the presence of about forty carved slabs decorating the outside wall. A few of the slabs depict objects that may represent astronomical sighting devices, but the major theme of the carvings is conquest, not astronomy. Each stone is inscribed with symbols and hieroglyphs that cannot be fully deciphered, but seem to represent the name of a place and the image of a vanquished ruler. A series of dates appears alongside these signs, presumably recording the time of battles or conquests. The dates refer to years, months, and days, and are set out in a double-column format typical of the writing of all ancient Mexican civilizations.

The Building J inscriptions do not, in fact, represent the earliest known writing or calendar dates known in the New World; at Monte Albán, hieroglyphs of this type, associated with carvings of grotesquely mutilated captives or slain leaders, date back to the first phase of construction on the mountaintop, around 500–400 B.C. But the decoration of an "observatory" structure, Building J, with symbols of conquest and chronology at least as early as 100 A.D. is highly significant. It indicates the aggressive political atmosphere and the emphasis on a sacred framework of time in which the observing skills of the ancient Mexicans were to grow in the centuries ahead.

Above, *inside the zenith tube built under the stairway of Building P at Monte Albán. In the photo taken on the second zenith passage day on August 5, 1976, the noon sun has penetrated through the opening of the shaft at the top of the picture. Below, early hieroglyphic writing from Monte Albán. The symbols on these upright stone slabs represent calendar dates and other undeciphered information, probably carved around 500-400 B.C.*

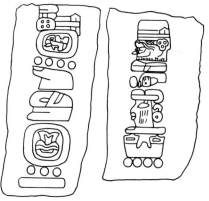

The early inscriptions at Monte Albán include calendar dates arranged in a system that seems to have had little connection with any obvious division of time. Days were referred to by pairing twenty different names with numbers from one to thirteen, so that they were known by such titles as 8 Deer, 5 Flower, and 13 Crocodile. The identical combination of twenty names and thirteen numbers could not repeat itself until 260 days had elapsed; on the 261st day, the cycle would begin all over again.

This 260-day calendar, which scholars call the Sacred Round, was vital not only to the people of ancient Oaxaca, but to all the civilizations of Central America. Indeed, it is still used for the timing of fiestas and rituals in remote highland villages today. Why should a period of 260 days have achieved such prominence in the minds of ancient Mexicans?

One possibility is that the 260-day cycle was based on the approximate length of human gestation. The interval between conception and birth was perhaps thought to be equivalent to nine lunar months, or nine female menstrual cycles ($9 \times 29 = 261$). This theory is especially plausible because of the custom of naming an infant after its birthday in the Sacred Round, which is still practiced in some rural native communities. The particular combination of name and number for that day represents a kind of astrological forecast, and is believed to influence the character and future of the child.

An alternative theory is related to the importance of the zenith passages of the sun in the tropical world. Is it possible that the calendar began at some place where the interval between the two dates when the sun shines overhead is exactly 260 days? This can happen only at a latitude of 15° north, which corresponds to the location of one important early ceremonial center. This is the site of Izapa, today nearly lost among the banana groves of the Pacific coastal lowland of Mexico close to the border with Guatemala. While Izapa consists of a series of once-impressive plazas with upright monoliths bearing carved reliefs of cosmological themes, no examples of early calendar dates have been found. There is no good reason for supposing that the whole calendar system emerged at one site and spread from there to all of Central America.

Indeed, it is difficult to identify any particular early culture as the most intellectually advanced in ancient Mexico. Although the first writing and calendar dates known at present are from Oaxaca, they were swiftly followed by inscriptions from regions as diverse as southern Guatemala, the Chiapas highlands, and the Gulf Coast of Mexico. Many different peoples seem to have participated in developing a basic system of writing and recording time which endured for many centuries. Even though the highest development of astronomical skills was to occur in the territory of the Maya, their knowledge was rooted far and wide throughout Central America.

A curious rock carving pattern demonstrates how cosmological and calendrical ideas were shared across great geographical distances. The basic design is simple, consisting of two axis lines that meet at right angles; this cross shape is usually enclosed by a pair of concentric circles. The design is not carved in a continuous line, but is instead "pecked out" as a series of dots, usually on the surfaces of natural outcrops commanding a clear view to the east.

The number of dots provides a hint of the cosmological significance of the pattern. In about a quarter of the carvings so far discovered, the total number of dots, or the sum of key portions of the pattern, adds up to 260 (in some cases, other important calendrical numbers such as 13 or 20 are present). Presumably, the central cross shape refers to the four directions of space, while the double circle may have a solar meaning, since in two examples a spiked fringe resembling rays of light fills the gap between each circle. While we cannot fully grasp the significance of this pattern, it seems safe to conclude that it was a sacred diagram of the universe, set within the framework of the "holy number" 260, which related to the flow of time.

While this expression of the unity of space and time must have had religious importance, the carving itself may also have had a secular purpose as a gaming board. According to the Spanish chroniclers, similar designs engraved or painted on the floors of Aztec houses were used for playing a popular game with counters called *patolli*.

So far, about forty of the cross-and-circle motifs have been identified in Central America, distributed across an astonishingly wide area from the steamy jungles of the Petén in Guatemala to the arid highlands of Durango in northwest Mexico. A full sixteen of these designs, however, are concentrated around one ancient site. This site is the great ruined city of Teotihuacán, which lies about one hour's drive northeast of Mexico City.

Today Teotihuacán is one of the leading tourist attractions in Mexico because of its two enormous pyramids, which were constructed around the first century A.D. The giant Pyramid of the Sun is just over two hundred feet high and probably contains more than thirty million cubic feet of earth, rubble, and mortar. While it is only half the height of Egypt's Great Pyramid, it covers roughly the same area at its base. The view from its summit overlooks a broad valley where several major routes converge, linking the Valley of Mexico to the Gulf Coast region.

Settlement began at this strategically important location around 200 B.C. The population grew swiftly, thriving on intensive cultivation practices in the valley and on special craft skills developed within the community itself (notably, the manufacture of tools from obsidian). By the first century, the power of Teotihuacán rivaled that of Monte Albán, which had risen to prominence at roughly the same time.

As an expression of their power, the rulers of Teotihuacán

SEEKING THE TROPIC: THE CARVED CIRCLES OF MEXICO

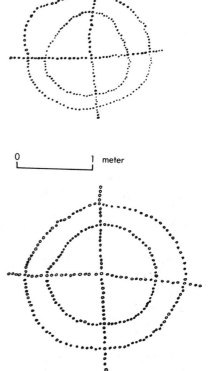

Typical cross-and-circle designs engraved at Cerro El Chapín, near Alta Vista, Mexico.

devised an ambitious, systematic new plan for the city. It was based around two great avenues intersecting at right angles, resembling the two axes of the cross-and-circle pattern. The two huge pyramids were positioned to reflect the layout of the avenues, and must have added greatly to the attraction of the city as a center of pomp and pilgrimage.

By the third century, the rulers were forced to accommodate a rapidly expanding population. They achieved this not by extending the city limits, but by organizing a program of "urban renewal." Their architects designed "apartment blocks" with standardized dimensions, each conforming strictly to the orientation of the city. It would have been easier to position the apartment complexes according to the natural topography, particularly in the outlying districts where space was plentiful, but instead the builders faithfully adhered to the original grid pattern over some twelve square miles of the city.

Why was this grid so compelling? Its origin is a mystery, for the major avenues run not due north–south or east–west, but appreciably "off" from the cardinal directions. (The so-called Street of the Dead is aligned at 15°28′ east of north, while the other avenue lies 16°30′ south of east). Several ingenious theories have been proposed to explain these deviations. No single solution is entirely convincing, but the cross-and-circle designs provide a strong clue. They were chipped laboriously into plastered temple floors within the city, and also into rock surfaces on the surrounding hills. The position of each carving, carefully surveyed by Anthony Aveni and his colleagues, suggests that they were not

View along the "Street of the Dead," the central avenue dividing the city of Teotihuacán, largest of all Mexico's ancient sites.

merely decorative, but helped the architects of Teotihuacán to establish the layout of their city.

To begin with, the markings suggest that the true cardinal directions *were* important to the builders; one cross-and-circle lies exactly due west of the center of the Sun Pyramid.

In addition, another pair of carvings may have marked the baseline that the architects used to fix the peculiar angle of the major "north–south" street. If a cross-and-circle design located near the Sun Pyramid is connected with another on the hill of Cerro Colorado nearly two miles away, the line forms an almost perfect right angle to the Street of the Dead. The purpose of this alignment, which may have determined the layout of the entire city, brings us back to the importance of the zenith passages of the sun.

During the first and second centuries A.D., when the grid plan was probably conceived, the Pleiades set just before dawn on the morning of the first zenith day. Viewed from the cross-and-circle engraving near the Sun Pyramid, the conspicuous cluster of stars would have disappeared in the brightening sky just above the site of the cross-and-circle engraved on Cerro Colorado. This alignment may have seemed particularly important because the path of the Pleiades also passed close to the zenith overhead at this season. If this explanation of the Teotihuacán plan is correct, then the symbolism of the cross-and-circle design was indeed powerful. It linked the all-important "shadowless day" with the harmonious design of the entire metropolis.

A sinister solar carving discovered in the ruins of Teotihuacán.

At the height of Teotihuacán's prosperity in the centuries after 250 A.D., the city population grew to 150,000 or more, larger than that of Imperial Rome. Trading and diplomatic activities connected it to distant corners of Central America. As the influence of Teotihuacán expanded, so the cross-and-circle design followed. It may have been a kind of sacred and political badge proclaiming the power of the city. The widespread appearance of the pattern may indicate that the Teotihuacán authorities were trying to influence the calendars and customs of their remote trading partners.

So far, no examples of the design have been found farther north than the Tropic of Cancer, some four hundred miles from the city. This location may have struck them as extraordinarily significant because of their preoccupation with the movements of the sun. At the Tropic of Cancer, it will be recalled, the sun shines in the zenith on only one day of the year—the day of the midsummer solstice. North of this latitude, of course, the sun never shines vertically overhead. The Teotihuacán authorities may have launched an effort to locate the Tropic. A pair of elaborate cross-and-circle designs almost identical to those at Teotihuacán was recently discovered by Charles Kelley, an archaeologist from the University of Southern Illinois. The designs are carved on a rock outcrop on the plateau of Cerro El Chapín, less than three miles south of the Tropic line at Zacatecas in

The twin solar alignments of Alta Vista.
Left, view from the ruins of the Sun Tem-
ple toward the Picacho peaks, where the
equinox sun rises to the right of peak B.
Right, view along the axis of a cross-and-
circle design carved at El Chapín; the sum-
mer solstice sun rises over peak B.

northern Mexico. Both carvings are arranged so that one of their axis lines points toward the prominent horizon mountain peaks known as Picacho. On the auspicious day of zenith passage at midsummer, the sun rises over the peaks when viewed from the site of the carvings.

Kelley believes that the cross-and-circle symbols were carved here not only to commemorate the discovery of the Tropic of Cancer, but also to fix the position of an important settlement nearby. About four miles north of El Chapín lie the ruins of Alta Vista, an archaeological site dating from around 450–650 A.D.; its public buildings are obviously modeled after the architecture of Teotihuacán. The main ceremonial structure, the Sun Temple, is arranged so that its corners point exactly due north, south, east, and west. If an observer stands in the Sun Temple and looks over the eastern corner, the equinox sun rises *over the same peaks of Picacho.* Kelley supposes that the observations carried out at El Chapín were then used to determine a propitious spot for the foundation of Alta Vista. His theory implies that the behavior of the sun at the Tropics was sufficiently well known and of such importance that it led to the establishment of an outpost of the Teotihuacán empire.

Even if this remarkable idea cannot be proved, there is no doubt that zenith astronomy figured prominently in the political and ceremonial affairs of the ancient Mexicans. While native peoples of the north were preoccupied with sky motions along the

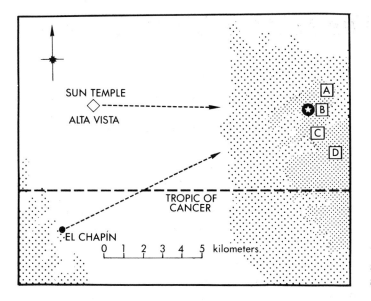

SUN TEMPLE

ALTA VISTA

A

B

C

D

TROPIC OF CANCER

EL CHAPÍN

0 1 2 3 4 5 kilometers

The twin alignments at Alta Visa, showing the Picacho peaks with letters corresponding to those on the preceding photos.

horizon (particularly with the solstices), the outlook of observers in the tropics was clearly influenced by the "up and down" horizon-to-zenith connection. This outlook was deeply rooted in prehistory, and lingered until very recently in the jungles of South America and on the islands of the Pacific.

Meanwhile the fortunes of the ancient Mexican empires rose and fell. Around the middle of the seventh century A.D., the Teotihuacán civilization collapsed, for reasons not yet satisfactorily explained by archaeologists. The city was devastated by fire, but it is not clear whether this was the only catastrophe or whether it was merely the aftermath of other events. With the demise of Teotihuacán, the Maya people far to the south entered a new age of prosperity and intellectual development. Over the succeeding centuries, the Mayan elite became the most skilled mathematicians and astronomers of the New World.

13. The Hidden Lines of Uxmal

THE PALACE OF VENUS

The ruins of the ancient Maya civilization astonished the six-teenth- and seventeenth-century European conquerors of Central America. A Franciscan friar named Diego López de Cogolludo noted their "magnificence" in the course of his history of the Yucatán peninsula, published in Madrid in 1688:

> When this land was discovered and conquered, there were found in it ruins that called forth powerful feelings of admiration in those authors having knowledge of them, as they continue to do today in those who behold what is left. They are manifold in the fields and on the wooded hillsides, some are vast structures especially those at Uxmal. . . .[57]

At this time, there was no agreement on who the constructors of the temples and palaces at sites such as Uxmal could have been, nor were legends preserved by the native Maya very helpful in understanding the problem. "We know not who they were, nor have the Indians it in their tradition" noted Cogolludo. He did, however, disagree with the opinion of many of his contempo-raries that the original builders were voyagers from high civiliza-tions on the other side of the Atlantic. Today our knowledge of the date and origins of the Mayan sites amply confirms Cogollu-do's viewpoint. The great cultures of the Old World—Babylon, Egypt, Greece, and Rome—had vanished long before building operations at Uxmal were at their height, probably around 800 A.D.

While Europe lay in the gloom of the Dark Ages, the skills of Mayan architects, artisans, astronomers, and mathematicians reached their peak. Nowhere are their achievements more obvi-ous than at Uxmal, now a popular tourist attraction only an hour's drive south of Mérida, the capital city of Yucatán.

In my visits to the site, I participated in survey work carried out during 1980 by a team from Colgate University led by Anthony Aveni and Horst Hartung. Aveni's original background was in astronomy, while Hartung is a professor of ancient architecture at the University of Guadalajara, Mexico. It took the collaboration of two experts from these separate disciplines to demonstrate that such distinct skills were in fact blended into one by the ancient Maya. The lines of their buildings were the product not merely of

One of the most graceful of all ancient Mayan buildings, the Palace of the Gover-nor at Uxmal, Yucatán.

a disciplined architectural style, but also of the demands of religion and astronomy.

We began our work at Uxmal by investigating one of the noblest of all surviving Mayan buildings, which the Spanish invaders named the Palace of the Governor. It is situated on three artificial stone-walled platforms, which raise the Palace above the level of the treetops and of the other structures at Uxmal. The long, low rectangular building stands on the highest and most commanding of these massive terraces.

A horizontal molding, running the entire three hundred feet of the facade, divides it into two parts, the lower severely plain and the upper ornamented with an elaborate stone mosaic. The contrast of these two parts of the façade conveys a graceful sense of balance. The effect is subtle, challenging the modern eye; it registered with particular impact on an early and eloquent observer of Mayan ruins, the American explorer John Lloyd Stephens, who asserted in 1841:

> There is no rudeness or barbarity in the design or proportions; on the contrary, the whole wears an air of architectural symmetry and grandeur; and as the stranger ascends the steps and casts a bewildered eye along its open and desolate doors, it is hard to believe that he sees before him the work of a race in whose epitaph, as written by historians, they are called ignorant of art, and said to have perished in the rudeness of savage life. [58]

The doorways open onto a double row of rectangular rooms. Like the interiors of most Mayan buildings, each is roofed by a corbeled vault and is now undecorated, dark and dismal, offering no clues to the activities that took place within. Only the decorated façade outside suggests that the structure was at least partly connected with celestial beliefs.

John Lloyd Stephens (1805–1852), whose travel books on Mayan ruins caused a popular sensation in the 1840s.

Anthony F. Aveni conducting fieldwork at Chichén Itźa, Yucatán, in 1980.

This upper façade is ingeniously assembled from hundreds of separately worked pieces of limestone. At first glance, it appears to consist solely of abstract shapes: an ornate framework of geometrical spirals, checkerboards, and crosses. Then one notices that certain designs form parts of a human face, a fierce, furrowed mask with a protruding nose, which is repeated again and again along the upper part of the Palace. A distinctive carved motif appears beneath each glaring eyeball: a pair of circles joined together by overarching curves which look like mustaches. Students of Mayan writing know that this is in fact a symbol for Venus, the two dots perhaps standing for the planet's dual aspect as Morning and Evening Star. We counted over 350 of these Venus signs under the eyeballs sculptured all over the façade. Whatever else this grim-visaged deity may have meant, its connection with Venus seems definite.

Further confirmation arose from the surveying work of Aveni and Hartung. Even a casual visitor may notice that the axis of the Palace of the Governor differs sharply from the orientation of most of the other buildings at Uxmal, which line up around 9° east of north. The long front of the Palace, however, faces 20° farther over to the east, a deviation which can be explained if sightings of Venus were carried out from its central doorway.

From the study of other Mayan sites and from illustrations in ancient manuscripts from central Mexico, it is known that the entrances of temples were sometimes used as astronomical observation points. With such comparisons in mind, Aveni and Hartung investigated what a priest would have seen from the central doorway of the Palace of the Governor. They found that every eight years, the Morning Star would rise at its farthest position to the south directly facing the doorway. In fact, when they accurately measured a perpendicular to the front of the Palace, it coincided to within one-thirtieth of a degree with this maximum southerly swing of the planet. A small bump on the level eastern horizon seemed to mark the spot.

A priest watching the horizon from a temple doorway, depicted in a manuscript from central Mexico.

Horst Hartung stands opposite the central doorway of the Palace of the Governor. View from the doorway, opposite, runs to the distant horizon, where Venus rises at its southern extreme every eight years over the pyramid of Nohpat.

Recalling the problems of megalithic astronomy, particularly the claims for highly accurate sight lines running to distant valleys or mountains, we might feel skeptical about this result. But, to their surprise, Aveni and Hartung discovered that the bump on the horizon was *not a natural feature*. After making their way along an arduous, overgrown path for several hours, they found that the "bump" was actually the top of a huge pyramid nearly one hundred feet high. This was the largest of the ruins at a site known as Nohpat, some six miles east of Uxmal. John Lloyd Stephens had stood on top of this same pyramid over a century ago and had admired the view, without realizing its astronomical interest.

> With the ruins of Nohpat at our feet, we looked out upon a great, desolate plain, studded with overgrown mounds . . . towards the west by north, startling by the grandeur of the buildings and their height above the plain, with no decay visible and at this distance seeming perfect as a living city, were the ruins of Uxmal. Fronting us was the Casa del Gobernador, apparently so near that we almost looked into its open doors. . . .[59]

Detail of the façade of the Palace of the Governor, showing the Venus symbol under the eyeball of the rain god mask.

It was at one of these "open doors" in the center of the Palace that the observers probably stood some twelve centuries ago, watching for the rise of the Morning Star over Nohpat. The three

alignments involved—the sight line to Nohpat, the perpendicular to the Palace, and the point on the horizon where Venus appears—all coincide so perfectly that this was surely a deliberately contrived arrangement. With so many sculptured Venus signs adorning the building, there can be little room for doubt.

So the positioning—as well as the decoration—of an important Mayan building seems to have been carefully designed with an astronomical motive in mind. Furthermore, the location of the Nohpat pyramid suggests that there may have been an element of conscious planning in the relationship between distant cities, again with astronomical ends in view. One startling theory proposes that the Maya conceived of their whole realm as bounded by four great cities at the north, south, east, and west. This idea is supported by carved inscriptions that record the names of these cities beneath particular signs for the four directions. Is it possible that the Maya conceived of sacred, celestially oriented territory on this grand scale?

At a less ambitious level, there were undoubtedly architectural practices that influenced the placing of one building relative to another within an individual site. Hartung has dedicated his career to the investigation of the curious hidden order which appears to underlie the plans of many Mayan sites.

Sometimes astronomy seems to have influenced these arrangements. For example, at Uxmal, a second alignment to Venus may exist, this time directed to its maximum southerly setting point on the horizon. The observer's viewing position was apparently high up on the tallest pyramid at Uxmal—the so-called House of the Magician—from a great doorway that is once again ornamented with the mask of a deity and symbols of the planet Venus. From there, the alignment passes over the exact centers of at least two other separate structures scattered across the site. This is admittedly less convincing than the case of the Palace of the Governor with its entire façade turned to face the rising point of Venus. But so many of these lines appear to exist at Mayan sites—usually connecting the centers of doorways, stairways, or courtyards—that we should at least consider them seriously, as Hartung has done in his painstaking research.

SECRETS OF THE NUNNERY

Hartung has presented many examples of these hidden architectural lines, but perhaps none is more curious and striking than that of the famous Plaza of the Nunnery at Uxmal. Here, four detached buildings, each with a superbly decorated façade that faces inwards, enclose a large open patio. This reminded early Spanish visitors of a convent; indeed, Father Cogolludo refers to the "nuns" of Uxmal and "a great court with many separate apartments in the form of a cloister where dwelt these maidens."[60] Today we can only guess at the actual function of the buildings, and admire their aesthetically satisfying arrangement. The visitor never feels claustrophobic in the courtyard, because all the buildings are raised on separate platforms at different levels. Each building has a different number of doors, and is set at a slightly different angle so that the patio is irregularly shaped. But it would be wrong to assume that there was no scheme underlying the layout of the Nunnery.

Hartung's plan, in fact, reveals a number of subtle relationships between buildings which a casual visitor would almost certainly miss, and which are best appreciated from studying the plan. It will be noticed, for example, that lines connecting certain doorways form a double set of axes that cross at almost perfect right angles near the center of the courtyard. The layout of the buildings seems to have been influenced by several lines running parallel to these central axes.

To put it simply, there appears to have been a disguised level of architectural planning at work behind the arrangement of these imposing buildings. This was a concealed order—perhaps with a sacred or magical meaning—that would not be obvious to the untrained eye.

Even details of the stonework contain hidden "messages" that have escaped notice until recently. In 1972 Weldon Lamb, an anthropologist from Tulane University, visited Uxmal for the first time. His curiosity was roused by the magnificent fretwork pattern adorning the front of the east building in the Nunnery courtyard. Was it merely decorative, or could it conceal an astronomical meaning?

There, above the central doorway and at the corners of the building, appears the grimacing mask of the deity once again, framed by the unmistakable double dots of the Venus symbol. The rest of the façade consists of a striking array of serpent heads which emerge from both ends of a series of horizontal bars, stacked one on top of the other. Until Lamb's visit, no one had

The Nunnery courtyard at Uxmal.

ever bothered to count the number of X-shaped pieces framed in the narrow gap between each of the serpent bars. There were thirty of these X's in the small group of bars over the central doorway, perhaps standing for lunar months. All the other X's total 584, which is an excellent average for the number of days Venus takes to orbit the sun. Lamb has noted other numbers connected with Venus that he thinks are represented by architectural features elsewhere in the Nunnery courtyard, but none of them are as remarkable as the total of 584 of the serpent-bar X's.

This figure is unlikely to be a product of pure chance. In one of the three or four surviving bark books of the Maya, the number 584 appears in the calculations of the planet's appearances and disappearances in the sky. Furthermore, the alignment of the Palace of the Governor to Nohpat indicates the far southern swing of Venus every eight years; this alone implies a long-term knowledge of the planet's movements. In addition, the hieroglyphs that appear on either side of the scowling, toothy masks seem to proclaim this to be "the Venus building."

The recent research at Uxmal shows that deliberate astronomical planning affected the Mayan architects in many different ways. Their celestial awareness seems to have influenced the positioning of buildings, not simply those located at a single site, but even ones separated by considerable distances. Intricate

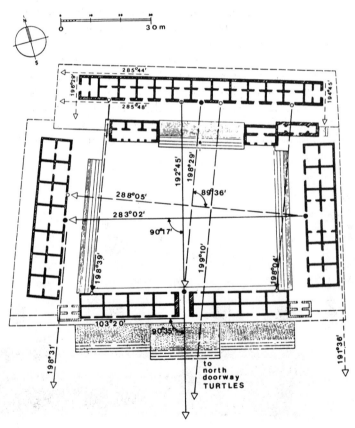

Survey of the Nunnery by Horst Hartung reveals intriguing architectural relationships, which are described in the text.

stonework, which one might suppose to be purely decorative, conveyed elaborate religious and astronomical meanings.

Even so, it may be wrong to assume that these buildings were designed for the exclusive use of devout astronomer-priests engaged in an idealistic quest for cosmic order and perfection. In fact, we remain ignorant of the exact purpose of most imposing Mayan stone structures. It takes little imagination to visualize the Nunnery courtyard as the residence of an aristocracy or priesthood, yet it could certainly have served a wide variety of functions. There may be a risk of exaggerating the influence of astronomy and religion in the planning of Mayan cities at the expense of other social, political, or economic factors.

What remains clear is that architectural capabilities were in no way separate from technical or ritual knowledge; such diverse skills were expressed as one. Some of these schemes—particularly of concealed numbers and orientations—are quite unfamiliar to Western architectural thought and practice, and so have escaped notice until recent years. Thanks to the recent period of interdisciplinary studies, we are only just beginning to realize the full complexity of the Mayan outlook.

A hidden Mayan message in stone. The east wing of the Nunnery displays this elaborate mosaic façade. The total number of X-shapes between the horizontal bars along the length of the east wing is the all-important Venus number, 584.

14. Beans and Crystals:
The Rites of the Modern Maya

THE CODE OF THE CHAMULAS

We read of the Maya so often as a "lost civilization" that it is easy to forget that the Maya are still alive today. In fact, they form the largest indigenous American group north of the Andes. Their population of some two and a half million is concentrated in northern Guatemala and the northern Yucatán peninsula, and also extends on either side into parts of Mexico (the states of Chiapas and Tabasco) and the western fringes of El Salvador and Honduras. The geographical isolation of many highland communities, as well as a history of bitter resistance to early Spanish conquerors and missionaries, ensures that the Maya cling tenaciously to their pagan rituals, although these are often blended with elements of Catholicism.

In fact, during the early years of the conquest, many fragments of Catholic doctrine were easily accommodated within the enduring Mayan system of beliefs. The symbol of the cross was identified with the four sacred directions of space. The story of the Crucifixion matched native traditions of human and animal sacrifice (in 1557, for instance, two young Mayan girls were actually crucified in a Yucatán village by priests inspired by a mixture of Christian and native teachings). The holiest of all forces, the sun, which was usually identified with male qualities, naturally became "Christ" or "Our Father," while the female moon was "Our Mother" or "The Virgin." In this merging of Christian and pagan traditions, we can perhaps catch a glimpse of how the ancient Maya regarded the universe and the passage of time, even though the image is distorted by Spanish influences.

The idea of an intimacy between the affairs of men and the movement of the sun is still basic to the beliefs of the modern

View of a Mayan hamlet near Chamula, Chiapas, Mexico.

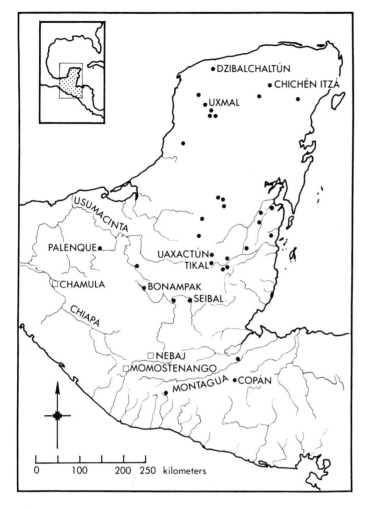

*Map of major ancient Mayan sites,
together with modern Mayan settlements
mentioned in this chapter.*

Maya. At San Juan Chamula, a cluster of hamlets scattered across
a rugged, forested plateau in the mountains of Chiapas, the forty
thousand Maya inhabitants consider themselves to be at the
center of the universe. Their church encloses what they think of
as the "navel of the earth." The high altitude of their village
places them in a superior relationship with the sun which no
other community is lucky enough to share. And the farther one
withdraws from the "navel" of Chamula, the more alien and
dangerous one's surroundings become.

Furthermore, time itself shrinks back on itself into the past as
one travels away from Chamula. So the outer fringes of the island
earth are populated by ancient demons, fierce animals, and, as
the Harvard anthropologist Gary Gossen found to his surprise, by
North Americans:

> Chamulas were fond of asking my wife or me the question, "Do
> people bite and eat one another in your country?" Surprised we
> would usually answer, "Of course not. Do people bite and eat each
> other here?" They would roar with laughter at our stupid question
> and then usually answer, "Well, no, but the first people did." Only

after participating in many of these seemingly absurd exchanges did the problem begin to make sense. I soon learned that the United States . . . lay in the outer, if not totally unknown, reaches of relative distance on earth. It followed as night the day that great social distance also pushed back the level of relative time, so that asocial behavior eliminated long ago in Chamula (such as infanticide and cannibalistic consumption of one's own children) might easily still occur at the outer limits of the universe.[61]

In the course of his research during the late 1960s, Gossen discovered that the intricate social code of the Chamulas is closely wedded to notions of time and space, and particularly to the progress of the sun as it sweeps out its path across the sky. Thus the four successive creations and destructions of the world are the consequence of the immoral conduct of earlier people and the Creator's efforts to achieve a better result each time.

Our present world (still part of the Fourth Creation) is a square island riddled with underground caves that lead to its outermost edges. The caves are inhabited by demons and also by the Earthlords, who are the masters of rain, clouds, thunder, and lightning. Beneath the earth is the underworld, where the dead eat flies and abstain from sex, but otherwise follow a normal existence. The sun passes through the underworld at night and emerges at dawn on the third and holiest layer of the sky. The face and head of "Our Father" are so hot and brilliant that the rays penetrate through the lower two layers to reach the earth, while the rest of his body remains invisible. The stars and moon (associated with the Virgin Mary) belong to the second layer. The circular paths of the sun and moon bind the whole of this layered creation together. The sacred orbits not only mark the limits of the island world, but also bring order and holiness to the cycles of day and night, dry and wet seasons, and years and generations.

A healing ceremony conducted at a shrine in Guatemala.

This vision of the universe is shared by many communities in highland Chiapas, with interesting variations from place to place. For instance, in the village of San Andrés Larrainzar, about nine miles north of Chamula, informants speak of the sky as a multi-tiered pyramid supported by a giant tree. Venus leads the way as the sun rides in a cart that climbs, and then descends, one of the thirteen flower-strewn steps of the pyramid each hour.

Should we be tempted to dismiss the Mayan outlook as naive and childlike, we should note that the simplicity of the sky models does not preclude accurate celestial knowledge and observations. Working far to the south of Chiapas in the mountains of Guatemala during 1974, an American anthropology student named Judith Remington interviewed Cakchiquel-Quiché Maya and recovered the names of many prominent stars and constellations, together with the seasonal and nightly timings of their appearance. The periods of visibility and invisibility of Venus seemed to be well understood by her informants.

This knowledge is not pursued for its own sake, nor are the Mayan concepts of the sky mere folktales told for amusement. Whatever the differences in detail, the cosmological schemes serve to order every aspect of social conduct.

In the midst of elaborate ceremonies, the Chamula priest will always pass counterclockwise to the right, because right-handedness is associated with holiness, good behavior, and seniority, as well as with male qualities and with the direction of north and with heat. Gossen explains this concept as the result of imagining the sun to be a person facing toward the west, with his daily path projected on the flat earth as an arc curving to the right. By contrast, the unlucky west and south are linked with death and the wet season, together with female qualities, inexperience, and coldness.

These clusters of qualities mean that life in Chamula is highly organized and patterned. (A baby and sick people are dangerously "cold," while the priest's ritual objects—his tobacco, rum, incense, candles, and fireworks—are holy because they emit heat. Women occupy the lefthand space in the house for sitting, eating, and working, while men occupy the right, and so on). The solar path provides the model for a tightly structured, conservative society.

For the modern Maya, all this is more than a matter of custom, since the remarkable fusion of space and time is even expressed in speech. A single word defines both the time and the place where the sun rises on the horizon. The ancient root word *kin* took on a variety of meanings, referring equally to "sun," "day," and "time." Its modern equivalent among the Chamula, *k'in*, significantly means "fiesta," the carefully timed religious ceremonies that mark the principal divisions of the year for ordinary villagers. Because of this preoccupation with the flow of time, a wristwatch is usually the most eagerly sought-after product of the outside world.

A Chamula healing ceremony photographed inside a shrine dedicated to the dark-skinned Saint, Nuestro Señor de Esquipilas, who is widely revered by the Maya. Soft-drink bottles are presented as offerings. Bottom, the priest-shaman binds a boy's head with a knotted cord representing the ropes that bind the universe, linking individuals to their animal souls in the sky. The priest prays to the Sun God (Christ) and to St. Jerome that the animal souls of the boy's family should remain whole, untouched by illness and disease.

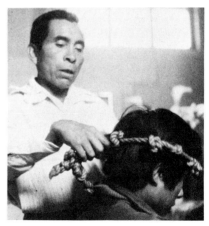

THE SACRED COUNT

It is highly likely that this special sense of time was characteristic of the ancient Maya, too, since even today their descendants preserve parts of the old calendar, which is still used in arranging the fiestas and foretelling the future.

According to a wide-ranging survey published in 1952, over eighty communities (mostly in Chiapas and Guatemala) were at that time still in possession of their calendars. Some towns had just the repeating cycle of 260 days, while farther to the north both the 260-day Sacred Round and the 365-day solar year were in use. The Sacred Round was associated mainly with casting fortunes for individuals, while the solar calendar ensured that the counting of the days could be kept in tune with the seasons and the planning of agriculture and public activities.

However, there was certainly no rigid distinction between the two calendar systems. Recent studies in highland Guatemala emphasize that the sacred and solar counts are closely interwoven in the Mayan mind, just as the private, fortune-telling aspects of the shaman-priest's work merge with his public duties to the community as a whole.

Why is this peculiar meshing of the solar year with an artificial cycle of 260 days so important to the Maya? The link between the 260-day period and the approximate length of human pregnancy is recognized by the Maya of several traditional communities, although a variety of other, less convincing explanations have been suggested by scholars in recent years.

The cycle is composed of thirteen numbers paired off with twenty named days of the month, so that exactly the same combination of numbers and days will repeat only after 260 days have elapsed. Most authors compare this to a picture of two gear wheels, one with thirteen teeth and the other with twenty, clicking off each day in turn. This mechanical image is utterly remote, however, from the way in which the Maya visualize the passing of time. The twenty days are not dull units, but gods whose particular aspects and associations determine the fortunes of that day. An elderly priest in highland Guatemala imagined their daily task as follows:

> The deity having the day's duties carries the sun in his chest, setting out at dawn on a litter supported by four lesser divinities. The sun shines forth from his chest as a great eye illuminating the earth. The god's headdress above is so large and magnificent that it requires six attendants. At the end of his journey he is sweating from the heat of the sun and the burden of carrying it. His attendants wipe the sweat from him as he gets out of his litter, exhausted from the day's work. [62]

The image of units of time as gods shouldering immense burdens is depicted in ancient Mayan art (for example, on the famous Stela D at the site of Copán). Today the day gods are still regarded with awe and reverence. An informant at Jacaltenango, Guatemala, explained that, after completion of the 260-day cycle, they should be thanked "as one should thank his parents

after eating."[63] The deified nature of the calendar, charged with personal significance, is clear from such accounts.

In 1982 one of the most complete descriptions of how the Mayan priests make use of this sanctified counting of days appeared in a book called *Time and the Highland Maya*. Its author, Barbara Tedlock, an anthropologist at New York State University, spent a total of about a year and a half living in one of the most staunchly traditional towns of highland Guatemala called Momostenango. In this community of forty thousand Quiché-speaking Maya, the shaman-priests are known as daykeepers (from the Quiché word *ajk'ij*, meaning day- [sun- or time-] keeper). They are so numerous and command such influence that they, not members of the local Catholic fraternities, control the running of the town's religious and political affairs.

The most unusual aspect of Barbara Tedlock's work was that she and her husband Dennis Tedlock were actually initiated as daykeepers during 1976. Their intense interest in the lore of the shaman-priests led them to be adopted as students by a prominent shaman, who also had favorable dreams connected with them. The Tedlocks impressed their Mayan teachers with their unusual proficiency in memorizing lengthy prayers and formulas (their efforts were in fact assisted by a tape recorder). Many hours of study spent with the priests have resulted in a remarkable "inside" view of Mayan beliefs and practices.

When a man experiences urgent, disturbing dreams, together with painful muscle cramps and other physical afflictions, these are believed to be signs that he should take up the vocation of daykeeper. If the symptoms persist, he seeks out a priest, who establishes by divination whether formal training as a daykeeper should begin.

Clients call on the priests for advice on virtually any topic, including the interpretation of dreams, the diagnosis of the sick, the character of a newborn child, and the best days on which to

This sculptured figure forms part of an elaborate inscription carved on Stela D at Copán. The figure represents the fourth of the nine Lords of the Night. On his back he carries the night, a jaguar skin whose spots are shown as stars.

buy and sell. The priest's job is not merely to predict the future, but also to influence the situation when affairs are not turning out well for the client. It is done with the arrangement of divining aids—usually beans and crystals—laid out on a table so that a sequence of days in the Sacred Round can be counted. If performed correctly, this operation causes the lords of the day to "speak" to the priest through his "blood." The "speaking of the blood" is described as tingling or twitching sensations in the blood or muscles, which occur in different parts of the priest's body and are interpreted by him according to their position. They lead him to suggest various remedies to the client such as offerings of prayer, incense, or liquor at the local shrines. In other words, the daykeeper's job combines the roles of astrologer, psychic, and doctor as we understand them in our society.

In some Guatemalan communities, the priests also regulate planting and harvesting activities with their calendars. At one village of the Ixil-speaking Maya, the farmers are reported to have worked quickly to finish the planting of a field on the right day. Just as the character of each particular day god provided a kind of astrological framework for the individual, influencing how and when he should conduct his affairs, so the workings of agriculture were set in divine order.

Uncorrected, repeating cycles of 365 and 260 days are not, of course, of much practical use; even the period of 365 is a quarter of a day short of the actual length of the year and will soon fall out of step with the seasons. But neither the ancient nor the modern Maya appear to have inserted a regular leap-year type of correction as we do. The timing of the calendars must have been fixed in some other way.

In Chamula, the 365-day calendar is adjusted each year by observation of the solstices, and is used extensively for the

The spring ceremony of K'in-Kurus conducted by a Mayan priest in a field near Chamula. The priest petitions the Earth Lords for rain as the wet season approaches around the beginning of May.

scheduling of agriculture. To be on the safe side, the dates of fiestas are checked with the modern Catholic calendar. Simple recording aids may also be in use to help them keep track of the solar year. In 1969 Gossen managed to obtain a wooden calendar board that had been kept by an elderly woman shaman in a Chamula village. She was reported to have risen each morning and made a charcoal mark for the day on the board, which she hung from a nail at the end of her bed; there were 365 marks in all. This tally, reading from left to right, was divided by a thick mark every twenty days, representing a new month. Alexander Marshack examined the calendar using infrared photography, and showed that the charcoal marks had been repeatedly erased and renewed, year after year. The tablet was said to have belonged to the woman's grandfather, so it could be over a century old.

The Chamula calendar board is the only example so far reported among the modern Maya of an attempt to record the passage of the year. However, other devices, including solar observatories, probably assist the priests in correcting the count of the days. A description of such an observatory was found among the notes on the Ixil Maya of Guatemala compiled over forty years ago by the American anthropologist J. Steward Lincoln. The Ixil occupy three townships situated among remote and spectacular high mountain ridges, usually shrouded in dripping mist which rises from the Petén jungles to the north (Lincoln died of pneumonia while working there in 1941).

Unfortunately, Lincoln left only a rough sketch of the observatory near the town of Nebaj, together with the comment that sunrise was watched there on 19 March 1940, two days before the spring equinox. It appears to have consisted of two separate alignments marked by upright stones, both sharing the same distant "foresight" stone and directed toward a notch on the horizon. To ensure accuracy, two observers probably watched the sunrise at the same time along different alignments. Other uprights—including a stone circle—are shown on Lincoln's sketch, but there is no information about how they were used. If similar observatories were once widespread, the highland calendars must have been accurately regulated and of direct usefulness to the ordinary farmer.

In fact, observatories remarkably like the stones of Nebaj seem to have existed long before the ancient Maya established their great cities. Far back in the so-called Pre-Classic period (perhaps as early as 800 B.C.), settlements in the highlands and south coast of Guatemala were arranged around courtyards featuring undecorated, upright stones. These plain stelae were placed in lines, or else erected against the bases or on top of earthen pyramid mounds. The archaeologist Edwin Shook has claimed that in certain cases the stelae probably mark solar orientations. At one

A wooden calendar board recently acquired in Chamula by Gary Gossen of the State University of New York. The names of the eighteen months of the traditional calendar (each twenty days long, with one extra five-day period) are noted on the board in charcoal.

THE SUN STONES OF THE MAYA

site called Finca Naranjo, an east-facing stone has a three-foot hole in the middle, apparently shaped so as to fit an observer's head and shoulders right inside.

Solar alignments also seem to have been commemorated by the Maya, notably at the famous site of Uaxactún, an extensive complex of buildings buried deep in the jungles of the Petén. In place of crude stelae and earth mounds, at Uaxactún an open plaza is surrounded by ceremonial structures of stone. From the top step of the pyramid at the west, one looks over to the opposite side of the plaza where there are three small, equally spaced buildings raised on top of a single great stone platform. Viewed from this step, the sun rises over the three small buildings at the key points in the year: at both solstices and equinoxes. This arrangement seems to have been deliberately copied in the architecture of at least a dozen other Mayan sites, although the pattern of astronomical alignments does not seem to have been followed consistently.

Other evidence for accurate solar observations may be present among the sculptured hieroglyphs of the ancient Maya. According to a theory published by the astronomer John Teeple in 1930, calculations relating the actual length of the year to the 365-day cycle and other time periods may exist, although this possibility has been strongly challenged by a number of other scholars.

Even if we discount the evidence of the true (or "tropical") year calculations, the ruins of the observatories suggest that an accurate solar reckoning was probably available at an early stage in the rise of the Central American civilizations. Through such observations, the ancient priests would have been able to adjust the 365-day cycle to the passage of the seasons, just as their descendants at Nebaj and Chamula did recently. So the calendar priests were not removed from everyday life. Then, as now, their knowledge

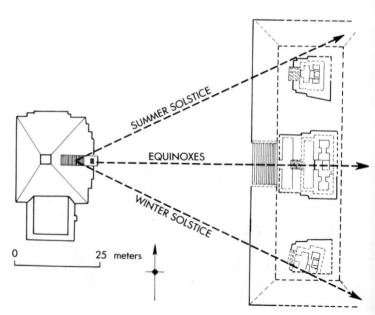

SUMMER SOLSTICE

EQUINOXES

WINTER SOLSTICE

0   25   meters

Plan of the Group E complex of buildings at Uaxactún, Guatemala, a solar observatory of the ancient Maya.

The ancient gods linger on among the Maya. Here, prayers are uttered before a monument dating to about 500–700 A.D. at El Baul, Guatemala.

must have been essential to the ordering of the agricultural routine and the succession of festivities and ceremonies.

Nevertheless, it would be wrong to imagine that the Mayan skywatcher was preoccupied with creating an efficient system for the management of earthly affairs. The evidence of their elaborate inscriptions and bark books proves that their concerns lay elsewhere. They seem to have been obsessed with the search for harmonious, resonating whole-number cycles—perhaps with the quest of one supreme whole number—which would divinely organize and unify both the heavens and human affairs. The correspondences between these sacred numbers were sought in calculations that often stretched far beyond a human lifetime.

The ancient Maya with his intricate computations carved in stone seems far removed from the Quiché *ajk'ij* counting out the days with his rows of beans and crystals. Yet even though centuries of violent historical upheaval divide these two figures, the modern Maya have clearly preserved more than just a bare practical framework for the ordering of their calendars.

An obsession with the flow of time seems to characterize the Maya of all periods. The peculiar attitude of the Chamula toward time and space helps us to imagine something of the outlook of the Classic Period astronomers. Just as the units of time invoked by the Quiché shamans are deities who affect the destinies of men, so we might expect the sacred calculations of the ancient Maya to have an astrological, fortune-telling aspect.

Remarkable advances in the study of the sculptured hieroglyphs achieved during the 1960s and 1970s have shown this to be true. In fact, the purpose of the inscriptions was never to record disinterested, intellectual games of mathematics. Instead, the intricately calculated time cycles were seen as omens and auguries that touched on the fate of individual rulers and their descendants. Nowhere has this fact emerged more clearly than in recent studies of the great complex of Mayan ruins at Palenque, in the state of Chiapas, Mexico.

15. Lords of Palenque

The most sensational discovery in Maya archaelogy took place in 1952, when the entrance to a secret chamber was found deep inside the largest temple at Palenque, an imposing stone structure known as the Temple of the Inscriptions. It was so named because of the great panels of hieroglyphic carvings embedded in the inner walls of the temple building that crowns the top of the steep, nine-layered pyramid. The hieroglyphs are heavily worn, but scholars have been able to use them to figure a probable dedication date for the temple of around 690 A.D.

In the spring of 1949, Alberto Ruz Lhuiller, the archaeologist in charge of digging and restoration at Palenque, was clearing debris from the floor inside the temple building when he noticed that the wall joints continued on down below floor level, hinting at the existence of a hidden chamber below. Furthermore, one of the huge floor slabs had a double row of holes bored into its surface, each sealed with a stone stopper, suggesting that the slab might be a removable cover. Once it was hauled out of position, workmen began exploring the entrance to a vault extending beneath the floor until, six feet down, the first step of a stairway was exposed. Access to the stairway was blocked by a mass of clay and rubble, which had to be hoisted up, one bucket load at a time, by the light of a gasoline lamp.

It took four seasons of arduous digging before the bottom of the passage was reached, no less than sixty-five feet below the temple floor. At last, a triangular upright slab was reached, obviously blocking an entrance. At its foot lay the discarded bones of six young people, at least one of whom was female. These were apparently sacrificial victims who had been of high-ranking status, to judge from their fashionably inlaid teeth. The skeletons hinted at the presence of something truly extraordinary beyond the triangular slab.

On June 15, 1952, Ruz Lhuiller succeeded in entering a burial crypt that had remained silent and intact for over twelve centuries. Understandably overcome by emotion, he found himself confronted by

> . . . an enormous empty room that appeared to be graven in ice, a kind of grotto whose walls and roof seemed to have been planed in perfect surfaces, or an abandoned chapel whose cupola was draped with curtains of stalactites, and from whose floor arose thick stalagmites like the dripping of a candle. The walls glistened like snow crystals and on them marched relief figures of great size. Almost the whole room was filled with the great slab top of an altar, on the side of which hieroglyphs painted in red might be distinguished, while on the upper surface only the fact that it was entirely carved could be made out.[64]

When the engraved and painted "altar," which weighed about

The Temple of the Inscriptions, Palenque,
photographed by Teobert Maler around the turn of the century.

The sarcophagus lid from the Temple of the Inscriptions, Palenque.

five tons, was raised up on automobile jacks, a sarcophagus was revealed underneath it. The grave was protected by a polished stone cover resembling the shape of a womb, and the walls of the cavity beneath this cover were stained red with cinnabar, perhaps to symbolize the color of west and the dying sun.

Here lay the decayed bones of a tall, well-built man, estimated to be about forty years old, who had evidently been in good health and had suffered no injuries. He was stretched out on his back, his torso and face covered with fragments of green jade jewelry, which stood out brilliantly against the red-colored stone

Key to the sarcophagus lid. A is for the upper jeweled dragon heads, B is for the middle, fleshed dragons, C is for the huge bony dragon heads of the underworld.

beneath. The most extraordinary artifact in the burial was a mask of jade mosaic that had covered the face, with staring eyes of inlaid shell and obsidian. The vigorous expression of the reconstructed mask conveys the power which must have been commanded by this man in the living world.

The discovery of the burial crypt at Palenque shed new light on the question of how Mayan society was governed. It was now obvious that wealthy individuals, the pinnacle of a ruling elite, were in charge. Additionally, the discovery proved that at least some Central American pyramids were intended to be burial structures, as their counterparts in Egypt had been over two thousand years earlier.

The beliefs of the Maya about their dead rulers are illuminated by the carvings and reliefs found in the tomb. The walls of the burial chamber are decorated with a procession of nine richly attired lords, sculptured in larger-than-life-size relief. They probably represent the nine lords of the night, the guardians of the underworld. The interior of the burial chamber may have been visualized as a setting for the afterlife. Indeed, the whole pyramid was built in nine layers or steps, perhaps as a model of the world below.

But it is the exuberant carvings on the sarcophagus lid that tell us most about the people of Palenque and their ideas of death. At first glance, it is difficult to pick out a pattern among the scrolls, crosses, and masks that envelop the central human figure. Do they really represent a space capsule with an astronaut aboard, as Erich von Däniken claimed in his best-selling book, *Chariots of the Gods?*:

> There sits a human being, with the upper part of his body bent forward like a racing motorcyclist; today any child would identify his vehicle as a rocket. It is pointed at the front, then changes to strangely grooved indentations like inlet ports, widens out, and terminates at the tail in a darting flame. . . .[65]

Agreed, the design *does* look like a rocket, and a child would enjoy the picture of alien space shuttles descending on the pyramids of Palenque. But a careful look at the carving suggests that this would be a rather odd space capsule. What, for instance, is a bird looking like an overstuffed rooster doing perched on the nose of the rocket?

More seriously, *all* the major elements of the sarcophagus lid design reappear on other carved panels at Palenque, but in combinations that bear no resemblance at all to space capsules. What do these compositions really depict?

IN THE JAWS OF THE SKY DRAGON

Beginning in 1973, an international conference of Mayan researchers has met each year at Palenque. The theories and discussions aired at these meetings have led to several breakthroughs in the deciphering of individual signs and glyphs, and to a growing understanding of the ideas expressed in the art of Palenque as a whole. The carvings on the sarcophagus lid provide a good example of the results of all this recent research, since the new interpretations of the engraved symbols offer a striking glimpse of the ancient Mayan vision of the universe.

We begin by noting that the central figure "reclines" in an awkward, unbalanced posture. Parts of his headdress are in disarray, suggesting that he may actually be falling through the air. Behind and above him rises a cross-shaped image representing the sacred plant or tree which the modern Maya still believe connects the underworld, the earth, and the heavens. But this cross is also probably meant to be viewed in a horizontal sense, too, as pointing to the four sacred directions of space. If this is correct, we should perhaps think of "up," the top of the composition, as representing not north but east, the key cardinal direction in Central America, while "down" stands for west. So the central figure faces the east, yet is falling toward the west, the place of sunset and death.

The cross is not just a vertical tree, then, but also a pivot of horizontal space. In this horizontal view, it represents the bodies of two double-headed dragons crossed over each other. The peculiar heads of the dragons with their long upper jaws can be seen forming the ends of each arm of the cross. All three heads are shown as if decked with jade necklaces and jewels.

Such dragons are among the commonest of all sky gods depicted in Mayan art. They were believed to generate rain at the four directions, and so were indirectly associated with fertility and crops. Symbols connected with the dragon apparently adorned the robes, headdresses, and scepters of the Mayan nobility, while many carved scenes were framed by bands that almost certainly represent the creature's body. The composition on the sarcophagus lid is itself enclosed by such a "dragon band" running all the way around the edges. This sets the entire design in a sacred, celestial frame of reference.

Returning to the main scene, we find that it displays not just one but *three* different sets of double-headed dragons. The body of the second dragon is draped in an inverted "U" around the branches of the tree, and ends, like the first dragon above, in grotesque, outward-facing heads. The final pair of outward-facing jaws stretch up from the bottom of the composition and almost embrace the human figure. They are barely recognizable, because the dragon has been reduced to a fleshless skeleton.

These three stacked layers of dragons surely stand for three levels of the universe. On top there are the jeweled dragons, probably a symbol of the heavens. In the middle are the fleshed dragons with their sinuous bodies, representing the forces of the living world. Finally, below, there are the skeletal creatures of the underworld; the central figure is literally falling into the jaws of death.

Many other images emphasize this idea of an ordered balance between the forces of life and death. The "rooster" at the top is a creature who probably had associations with both the sky and the underworld, as the "sky serpent" who heralds the dawn, and also as the screech owl who rules the night.

Meanwhile, at the bottom, a grotesque toothy head stares out between the dragon bones. The mouth and lower jaw are all bone, too, but the rest of the head remains fleshed. A badge emblazoned on the monster's head provides the clue to its identity. The X-shaped emblem indicates the Mayan root word *kin*, which, as explained in chapter 14, stands for time, day, and sun, all combined. It seems very likely that the monster as a whole is meant to represent the sun. Along with the human figure, the sun is falling into the dragon jaws of the lower world; in other words, it is sinking beneath the horizon at the moment of sunset. The sun's embrace by the jaws of death could also refer to the winter solstice, the low point of the sun's annual cycle.

The composition on the sarcophagus lid strikes us as a cluttered, confusing collection of symbols. Yet there is an underlying theme, a unique Mayan vision of the world, uniting the whole scene. All of these symbols combine opposites in one—profile faces are hidden in frontal heads; vertical trees are also horizontal dragons; life and sky signs always balance those of death and the underworld. The human figure himself is poised between these opposite states, at the instant of transition from the living world to the realm of the dead. The fate of the human is not merely involved in this grand cycle of forces, but is clearly identified with the path of the sun itself. This suggests a divine, superhuman status for the man who is depicted in his moment of death.

One final image deserves special comment. At the same level as the head of the human figure, but on the opposite side of the tree trunk, an isolated symbol appears in the form of a combined cross and rosette. This sign is usually employed in the inscriptions to indicate "completion" at the end of a sacred calendar cycle. Its presence here, balancing the head of the human, must be significant. The man is dying at an appropriate time, completing the destiny of the heavens and of his ancestors as a sacred unit of the calendar elapses.

The moment of passage to the underworld is not fearful and chaotic, then, but filled with holy purpose and precise orderliness. The entire design on the sarcophagus lid asserts the importance and divine status of the individual within the framework of nature, intimating that the Maya placed man literally at the center of things. They identified the human cycle with those of the heavens and of all other natural forces.

Was there a sense of faith, then, in the survival of the individual after death? One of the clues offered by the carving is a curious sign shaped like a bone, which touches the tip of the human figure's nose. The Huichol people of northern Mexico today believe that the soul is contained in the bones and is reborn from them, so this little bone may represent the soul or life force emerging on the breath. The efforts involved in carving the lid, not to mention the construction of the pyramid itself, may all have constituted a kind of prayer for the dead man, either for his rebirth in the lower world or for his transformation to a god in the upper.

The burials found underneath many temples were probably not just incidental acts of piety. Instead, a cult of ancestor worship seems to have infused the whole idea of the pyramid. The presence of the burial deep inside the Temple of the Inscriptions remained important long after access to the crypt had been blocked by hundreds of tons of clay and rubble. The stairway

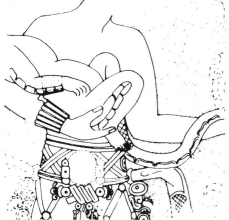

The front piers of the Temple of the Inscriptions at Palenque bear life-size reliefs of a woman holding a divine infant. Like many other reliefs at Palenque, it has been seriously damaged by recent acid rain.

leading up from the mortuary chamber was arranged so that a small stone tube of square cross section ran all the way up its right-hand side. This was surely to preserve an oracle or channel of communication with the dead man, or else to release the spiritual energies that may have been thought to flow from his bones.

The little stone tube did not stop once it had reached the temple floor. Instead, its course seems to have continued, or at least to have been marked symbolically, by a line of inset stones that runs right across the temple floor. This line emerges on the outside, at the base of one of the four great piers that support the front of the temple. Each of these piers is ornamented with a panel of stucco relief, depicting a life-size figure holding a baby or a small child.

So the supernatural energy tapped by the stone channel seems to have been linked to one of the four great images of infancy that dominate the front façade of the temple. We might reasonably conclude that this was done to illustrate a general religious belief about rebirth, perhaps even a Mayan "Nativity" legend. Actually, recent research has revealed startlingly precise information about the meaning of this imagery. We can now pinpoint the identity of the baby, the buried man in the crypt, and many of the greatest rulers of Palenque.

THE LEGACY OF LORD PACAL

This breakthrough was largely accomplished in the early 1970s by such scholars as Floyd Lounsbury of Yale University and Linda Schele of South Alabama. They pursued earlier ideas that the dates of birth, death, accession, and other events in the lives of individual rulers could all be recognized among the inscriptions at Palenque.

Important clues for this research were found among the carvings that adorn the sides and ends of the great sarcophagus in the Temple of the Inscriptions. These carvings are clearly portraits of individuals, who are shown emerging from the ground, crowned with foliage as if they were sprouting plants. Each portrait is

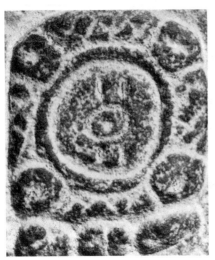

Glyph representing Pacal's name.

accompanied by signs, or glyphs, that can be shown to represent the person's name. We actually know what some names sounded like, as a result of careful comparisons with modern Mayan dialects.

It has been established beyond reasonable doubt that the man buried in the sarcophagus was called Pacal; his full title was Lord Shield-Pacal, to judge from the shield sign accompanying the glyphs, which can be read phonetically.

Who was this remarkable man? The history of his royal ancestors and immediate relatives can be traced in some detail, for they are the figures on the sides and ends of the sarcophagus, depicted as if they were plants. The inscriptions recorded on many different carved tablets at Palenque can be assembled to supply essential dates connected with their lives, though so far it has proved difficult to reconstruct actual historical facts or circumstances.

From this evidence, we know that women as well as men held power at Palenque. Pacal's mother, Lady Zac-Kuk, was one such leader, and it is likely that she acted as a regent when Pacal first assumed office at the age of twelve and a half.

According to the hieroglyphs, Pacal's rule lasted until his death at the age of eighty years and five months, probably in the year 683 A.D. During this period, Palenque grew to a position of greatness among the western settlements of the Maya. Its influence was reflected in widespread trading contacts that reached as far as the site of Copán in present-day Honduras.

The majesty of Pacal, ruler of Palenque, is conveyed by this stucco relief ornamenting the front of the Palace.

Pacal ordered the construction of the major buildings and the magnificent Palace at Palenque, and had the main plaza laid out in front of it. On the adjoining side, his architects then began their most ambitious project, the Temple of the Inscriptions, conceived as their ruler's tomb and the setting for his transformation to a god. The pyramid was probably unfinished at the time of his death.

Pacal must have been an exceptional, revered leader, because each of his successors lavished a great deal of effort on proclaiming their relationship to him. His son, Lord Chan-Bahlum, came to power at the age of forty-eight, and was apparently responsible for completing the temple on top of the pyramid. According to an ingenious theory of Merle Greene Robertson, a researcher who lives near the ruins and has devoted many years to recording what remains of the sculptured reliefs, Chan-Bahlum probably ordered the decoration of the four front piers of the temple. The child depicted on each of the piers seems to be Chan-Bahlum himself, represented as a divine infant, and the life-size figures were probably intended to proclaim his authority as the rightful heir to Pacal. As if to emphasize this, the symbolic stone channel ran from the base of one of the carved piers all the way down to his father's bones buried far below. Perhaps the tube was imagined as an "umbilical cord" connecting the generations.

No doubt it would have seemed disrespectful had Chan-Bahlum attempted to outdo his father by erecting an even more magnificent pyramid. Instead, his architects hit upon a highly original

View of the Palace at Palenque.

plan for three temples, known today as the Cross Group. They were artfully arranged around the sides of an intimate plaza, positioned back against the steep hills at the southeast corner of the site. Each of the three temple buildings encloses a small sanctuary with ornately carved panels at the entrance and the back wall. John Lloyd Stephens, who visited the Cross Group in 1839, recorded his awe and perplexity at the sight of these carvings:

> We could not but regard it as a holy place, dedicated to the gods, and consecrated by the religious observances of a lost and unknown people. . . . Lonely, deserted, and without any worshippers at its shrine, the figures and characters are distinct as when the people who revered it went up to pay their adorations before it. To us it was all a mystery; silent, defying the most scrutinizing gaze and reach of intellect. Even our friends the padres could make nothing of it.[66]

Stephens would be astonished today if he could know the extent of our current understanding of the detailed symbolism and time intervals expressed on the sculptured slabs.

We know that the three temples of the Cross Group were dedicated in 692 A.D. as a monument to Chan-Bahlum's inauguration. Each of the main decorated panels displays a full-length profile of the new ruler at the moment of receiving the regalia of office from his father, Pacal, who stands on the opposite side to him. Pacal, of course, was dead by this time, and his supernatural state seems to be indicated by his shrunken stature in comparison to his son.

The special religious meaning and rituals associated with each shrine are presented by the elaborate symbols, resembling icons, between the two royal figures. Though the details of the designs differ, they all express common themes. Widespread beliefs springing from the seasonal, agricultural round blend with symbols of the exclusive ancestor cult that grew up around the godlike Pacal. Each temple seems to commemorate a special moment in the annual cycle of rituals, as well as a different aspect of the royal lineage.

The meaning of the engraved scenes was apparently reinforced by astronomical alignments of an imprecise kind. The Temple of the Sun, for example, bears an image of the sun sinking toward the underworld similar to the image on the sarcophagus lid in Pacal's tomb. The front of the building roughly faces the direction of midwinter sunrise, perhaps associated in the builders' minds with the symbolism of the carving.

Chan-Bahlum also seems to have displayed respect for Pacal by positioning the Temple of the Cross in a significant astronomical relationship with his dead father's tomb. The doorway lines up with the distant front façade of the Temple of the Inscriptions (this is the façade ornamented with life-size reliefs that are thought to depict Chan-Bahlum as a divine infant, rightful heir

A line running eastward from the front façade of the Temple of the Inscriptions, foreground, meets the central doorways of the Temple of the Cross, background. In the reverse direction this line points roughly to the midsummer sunset.

The Cross Group at Palenque.

Relief from the Temple of the Cross depicts Chan-Bahlum at the moment of his succession in 692 A.D., holding a ceremonial scepter.

to the lineage of Pacal). Viewed from the Temple of the Cross, the direction of the façade is roughly where the sun sets at midsummer. A few other astronomical alignments, all of this imprecise kind, have been noted at Palenque by such scholars as Anthony Aveni, Horst Hartung, and John Carlson.

So the unusual layout of the Cross Group corresponded to a remarkable combination of symbolic ideas, all adding to the pomp and sanctity which Chan-Bahlum hoped would be associated with the start of his reign. He enjoyed another twelve years of power until his death at the age of sixty-six.

The hieroglyphs record the names and dates of several of his successors, some of whom made notable additions to the architecture of Palenque. For example, a unique four-story tower, which lends a curious pagodalike air to the great Palace, may have been constructed in the rule of Lord Kan Xul, Chan-Bahlum's heir. It has often been suggested that the tower was built as a vantage point for the astronomer-priests of Palenque. Today it is the only manmade structure from which one can secure an uninterrupted view of the level northern horizon, stretching a clear eighty miles all the way to the Gulf of Mexico. Unfortunately, no inscriptions survive to confirm the observatory theory, except for a large, solitary glyph representing Venus, painted rather crudely in red on a wall above an inner stairway of the tower.

MAYAN MYTHS AND "MAGIC NUMBERS"

Whatever their special contributions to Palenque, all of these later rulers shared one particular concern: to relate their credentials back to the reign of Pacal and other ancestors by means of complicated calculations. The superb quality of many sculptured tablets suggests the prestige that was at stake in these mathematical and astronomical claims to power. A disturbing question naturally arises: do these records represent history? Or was there a temptation to falsify the timing of events to bring them in line with some favorable cycle of the calendar or the planets?

Suspicions are aroused, first of all, by the ages credited to the rulers at Palenque. The inscriptions inform us that Lord Hok died at seventy-six, Chan-Bahlum at sixty-six, while three separate carvings put Pacal's death at the venerable age of eighty years and 158 days. No doubt these dignitaries led a pampered existence, protected from the risk of disease and infection that must have limited the lifespan of the Maya commoner. Yet the records of Pacal's death clash with the evidence of his skeleton recovered from the great sarcophagus. Although much decayed, his bones indicate that he was only about forty at the time of his death. This mystery remains unsolved at the moment.

In any case, the task of the Mayan scribe was certainly not to document everyday events with a sober, unprejudiced eye. For instance, experts have long puzzled over the interpretation of the huge panels of hieroglyphs located inside the Temple of the Inscriptions. Although many of the symbols still cannot be deciphered, certain passages undoubtedly refer to both deities and historical humans. Events in the lifespan of these mortal and supernatural beings are related to immense calculations which go backward and forward, stretching over one million years into the past and over four thousand years into the future!

What was the purpose of such far-ranging computations? A study by Floyd Lounsbury published in 1976 delves into the remarkable mathematics that seem to have influenced the choice of an important inscribed date: the foundation date carved on the Temple of the Cross at Palenque.

The usual Mayan method of commemorating historical events was to cite the interval that had elapsed since the beginning of the present creation: a starting point far back in time, which can be set very tentatively in our own chronology at 3113 B.C. In a sense, this date was the equivalent of our 1 A.D., the base point from which all Mayan history was usually reckoned.

What is so peculiar about the inscribed date on the Temple of the Cross is that it was worked out from a slightly *different* base point a little before the beginning of the present world: in our terms, around 3120 B.C. Lounsbury was able to show that this was probably the result of fixing a date back in the previous creation which would share the same number properties as Pacal's birthday in the present world. In other words, this special zero point was the exact equivalent of the auspicious day of Pacal's birth, projected back into mythological time.

Red-painted Venus glyph in the Tower at Palenque.

The Temple of the Cross was no ordinary building, and a significant moment for its dedication had to be chosen. What if many different celestial numbers coincided at the time of its dedication, reckoned from Pacal's ancient birthday? This would surely invest the temple and its founder, Chan-Bahlum, with holiness and supernatural power. Was the interval of days separating Pacal's mythical birth and the dedication of the temple—1,359,540—some sort of "lucky number"?

Lounsbury discovered that this number can be divided exactly by no less than *seven* important calendar or planetary cycles. This property must have struck the priests as highly significant indeed. To put it simply, ingenious numerology was set to work to determine the exact day when the carving should be dedicated. Whether a ceremony of dedication actually *did* take place on that date, we cannot, of course, be certain.

There is a bizarre twist to the Mayan number game so brilliantly exposed by Lounsbury. The priests had contrived a day in the previous creation that would correspond to Pacal's actual birth, yet they did not think of it as "Pacal's legendary birthday." Instead, the glyphs identify the occasion with *the accession of a woman*, a mythical ancestress or ancestral mother, who was probably a celestial deity. Her name is indicated by a rather unflattering symbol in the shape of a grotesque animal head with an upturned nose. The inscriptions state unambiguously that Lady-Beast-with-the-Upturned-Snout was *over eight hundred years old* at the time of her accession.

Who dreamed up this strange myth of a fantastically aged goddess? Could it have been Pacal himself, who may have wished to emphasize the importance of his own birth by relating it to an event far back in the life of an imaginary ancestress?

This example from the recent flowering of research at Palenque illustrates the peculiar nature of the hieroglyphic writings that have survived from the Classic Maya period. Most inscriptions were intended to justify the activities of a ruling family in terms of their godlike ancestors. This was often done with the help of time intervals that appeared significant because they combined several astronomical cycles at once. These "magic numbers" were symbols of a divinely ordered framework of time and space.

It seems very likely that the priests would have delayed the holding of important ceremonies until the numerologically favorable moment had arrived. We can imagine how their calculations might have affected the timing of many lesser events besides inaugurations, such as royal marriages and military expeditions.

In fact, a 1982 study by Lounsbury and Mary Miller, then a graduate student at Yale, shows how an astronomical date was probably selected for a battle depicted in a famous mural preserved inside a temple building at Bonampak, Chiapas. To summarize their conclusions crudely, it seems that the battle (which probably took place in the late eighth century A.D.) was

Shell carving of a priest at Palenque.

A modern copy of one of the ancient Mayan murals discovered in 1946 at Bonampak, Chiapas, Mexico. This shows the aftermath of a battle as the victorious warrior–king presides over the humiliation of prisoners.

timed to coincide with a special appearance of Venus in conjunction with the sun and a prominent constellation.

On the other hand, it will never be certain whether the surviving inscriptions are accurate reflections of history, since the urge to meddle with the true dates to prove an astrological point may have been overwhelming. The carved panels on the Temple of the Cross are probably one such example of a "propaganda" document, as Floyd Lounsbury suspects:

> That there are inconsistencies in the inscribed record of the Temple of the Cross, such as might have been due to conflicting traditions or, as suggested here, to deliberate tampering with the record, is well known to those who have puzzled over it.[67]

Above all, we can be sure that there was no such thing as "pure science" among the Maya of Palenque. This is not to deny the existence of expert astronomers and mathematicians at the site, but their skills were obviously inseparable from the "divine politics" exercised by the successive lords of Palenque. Both astronomy and history were manipulated in the service of an ancestor cult which upheld the interests of the ruling family. Today we can scarcely begin to appreciate how powerfully these ancestor beliefs must have haunted the minds of the devout Maya, convinced as they were that their dead rulers were not lost to the past, but were gods who influenced the world of the living.

16. Prophecy and Precision: The Written Evidence for Mayan Astronomy

Our understanding of the hieroglyphic carvings at Palenque and other Mayan sites has improved dramatically in recent years. We now have glimpses of the religious and political forces that shaped Mayan astronomy—notably, the potent beliefs in supernatural ancestors and their ability to influence the fate of living leaders. The quality of the spiritual outlook of the Maya is so unfamiliar to us that it is easy to dismiss them as a thoroughly irrational people. If we do so, however, we ignore the solid intellectual achievements of their skywatchers, particularly their skillful prediction of lunar and planetary cycles.

The precision of the observations documented in the few surviving hieroglyphic books is astonishing. For instance, one book contains a scheme for the correction of Venus observations which ensures an accuracy of approximately *two hours in five hundred years*. Who were the officials responsible for these remarkable records? How were they able to score such phenomenal success in their observations?

The best-preserved of all the written manuscripts is the so-called Dresden Codex, which was acquired by the library of the German town in 1740. This picture book was compiled sometime after 1000 A.D. in the northern Yucatán, but there is evidence that it was based on earlier astronomical sources, and indeed it was painted over and rewritten at least once. A fine brush dipped in red and black vegetable dyes was used to draw the hieroglyphs; additional blues and yellows were added to the pictures. The parchment was made from flattened bark and treated with a thin, glossy coating of lime. It was assembled in the form of rectangular pages, joined together at both edges like the folds of an accordion.

While the Dresden Codex is the key source for understanding Mayan "science," it was neither a mathematical table nor an astronomical manual, but, rather, an almanac probably designed to assist the priests with divinations and prophecies. Astronomical and calendrical information appears alongside lively, cartoonlike images showing the dire omens associated with particular celestial events. Once again we encounter Mayan skywatching skills dedicated to the purposes of ritual prediction and astrology.

Unfortunately, only three other Mayan bark books besides the Dresden survive (and one of these three, the Grolier Codex, discovered in 1971, is controversial, though the majority of scholars now regard it as authentic). If we try to imagine what a future world would make of Western science with only a half

THE BURNING OF THE BOOKS

A Venus almanac from the Dresden Codex.

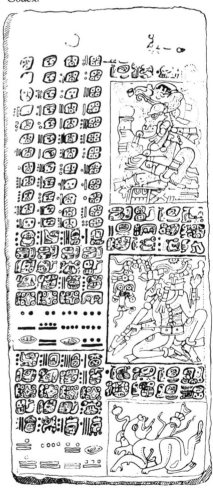

Friar Diego de Landa (1542–1579).

dozen university press books left over from a holocaust, the problem of assessing Mayan knowledge is obvious.

The Mayan libraries that may once have existed were systematically destroyed by Spanish friars in the mid-sixteenth century. Their actions were part of a repressive campaign to exterminate pagan beliefs, and involved all the apparatus of the Inquisition, including flogging, torture, and imprisonment.

One July day in 1562, Friar Diego de Landa ordered the burning of five thousand "idols" and twenty-seven hieroglyphic books in the village of Maní in the Yucatán: " . . . since they contained nothing but superstitions and falsehoods of the devil we burned them all, which they took most grievously, and which gave them great pain."[68] Ironically, having done his best to wipe out native learning, Friar Landa was responsible for passing down more knowledge of every aspect of Mayan society—including the specific meaning of certain hieroglyphs—than any other Spanish chronicler. Soon after the book-burning episode, Landa was removed from office and sent back to Spain to face trial for his excessive cruelties. In the course of lengthy examinations which resulted in his eventual pardon, he wrote a volume in his defense in 1566, his *Relación de las Cosas de Yucatán*. Even today, the *Relación* provides a detailed and fascinating account of sixteenth-century Mayan life.

In Landa's text, we learn of the native priests whose main attention, Landa says, was devoted to the sciences. The high priest was titled *Ahkin May*, meaning "He of the Sun," suggesting the importance of his astronomical duties. This was a hereditary male office, treated with great reverence by the secular lords who presented him with gifts and offerings. He presided over the appointment of lesser priests, examined them in their ritual knowledge, and provided them with books. In turn, they were responsible for instructing religious novices, who were chosen from the sons of other priests and the second sons of lords. The sciences they taught, explains Landa,

> . . . were the reckoning of the years, months and days, the festivals and ceremonies, the administration of their sacraments,

The Grolier Codex, below, one of only four surviving Maya books, was reportedly found in a cave in Chiapas around 1966, and was first described by archaeologist Michael D. Coe in 1971. Recent examinations by John Carlson strongly support the authenticity of this Venus almanac, probably dating to the 13th century A.D.

the omens of the days, their methods of divination and prophecies, events, remedies for sickness, antiquities, and the art of reading and writing by their letters and the characters wherewith they wrote, and by pictures that illustrated the writings.[69]

These duties were supplemented by the work of other religious men, such as the *chilánes* or shamans, who treated illnesses and uttered oracles in the midst of narcotic trances. Landa describes at great length, and with obvious disgust, how the priests organized ceremonies of human sacrifice which were, however, rare events, "on occasions of great tribulation or need."

Unfortunately, we cannot project this picture of the priesthood directly back to the period of Classic Mayan civilization. Major social upheavals had intervened in the Yucatán following the arrival of aggressive, militaristic factions from central Mexico in the period around 1000 A.D. But in view of the ritualistic, astrological significance of many earlier Mayan inscriptions, it does seem likely that a priesthood of some kind had always been in charge of compiling astronomical records. It may well have been a formally organized body similar to that described by Landa, since the almanacs preserved in the bark books were based on decades, if not centuries, of systematically accumulated observations.

This grim relief from the Tzompantli, or Skull Rack, at Chichén Itzá suggests the importance of human sacrifice among the later civilizations of Central America. The 15th century Aztecs, however, were the only group known to have practiced sacrifice on a massive scale.

The numbers and pictures in the bark books have been intensively studied ever since 1880, when the Dresden librarian, Ernst Förstemann, announced his discovery of their astronomical significance. A full description of these complicated almanacs would quickly exhaust the patience of an ordinary reader. It is not difficult, though, to grasp a general idea of how the Maya approached the problem of prediction. This understanding is

MAYAN METHODS OF PREDICTION

vital to an appreciation of the intellectual achievement of their astronomers.

At the core of all the astronomical texts was a concern for the Sacred Round of 260 days. As we have seen, this was an endlessly repeating chain of days with no obvious relationship to any celestial rhythm, yet it was indispensable to the divinations of the priests. Whatever the subject of the text, it was always arranged to bring the reader back to the exact same starting day in the 260-day count. This in itself called for considerable ingenuity in handling numbers. It also represents a basic difference from Western habits of understanding celestial events. The emphasis of the almanacs was *not* on when the predicted event would occur in relation to the seasonal rhythm of the sun. Instead, their purpose was to keep the event perfectly in line with the sequence of 260 days.

Besides the tables specifically devoted to the forecasting of eclipses and Venus motions, the Dresden Codex contains many other 260-day almanacs related to topics such as rainmaking, fertility, and evil omens, all matters of concern to the Mayan soothsayer.

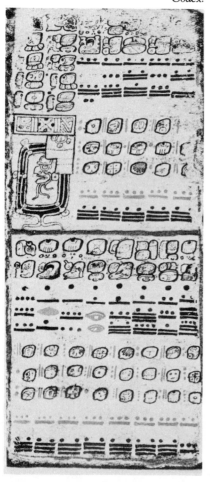

Lunar eclipse table from the Dresden Codex.

They must have been particularly worried by eclipses. In the sixteenth century, the Spanish chronicler Sahagún described the pandemonium that broke out among the Aztec on the occasion of an eclipse. When the moon grew dark, pregnant women put pieces of obsidian between their lips or breasts to ward off horrible birth defects, including the possibility that they would conceive mice instead of humans.

Though it was a rare event, a solar eclipse provoked widespread terror. Captives were sacrificed, while everyone drew their own blood. Sahagún writes:

> All were disquieted, unnerved, frightened. Then there was weeping. The common folk raised a cup, lifting their voices, making a great din, calling out, shrieking. . . . And in all the temples there was the singing of fitting chants; there was an uproar; there were war cries. It was thus said "If the eclipse of the sun is complete, it will be dark forever! The demons of darkness will come down; they will eat men!"[70]

Such fears may have driven the Mayan priests to record the dates of lunar eclipses over a number of decades. They would have been particularly concerned with the interval of time (usually about six months) that elapsed between each alarming event.

The interval is not *exactly* six months, because eclipses are governed by the complicated combined effects of the movements of the sun and moon. The relationship between the two motions results in a period of 173 1/3 days for the sun's apparent journey between one danger point and the next. This interval is a little less than the 177 days of six complete lunar months (and is called by astronomers the "eclipse half-year"). There is evidence to suggest that the Maya were actually aware of the 173 1/3-day period, but because they had no use for fractions or the kind of geometrical thinking that underlies our own understanding of these motions, they developed a highly original method of prediction.

Eight pages of the Dresden Codex are filled with a table that is clearly devoted to the timing of lunar eclipses. One column of numbers consists mostly of the figure 177 repeated over and over again, while another column allows the priest to keep track of the steadily accumulating sum totals. Every now and then, however, the figure 148 intrudes among the monotonous recital of 177s in the first column.

The number 148 is significant because it equals the number of days in *five* lunar months. In other words, the composer of the table was trying to adjust his record of six-month intervals to fit the real pattern of eclipses by periodically inserting a shorter, five-month period. In this way, the sequence of numbers allowed the priests to keep track of the "eclipse half-year."

What were they trying to do with this ingenious table? The document may have been intended partly as a kind of "diary," arranged so that the dates of actual observed eclipses could be accurately recorded in the 260-day calendar. However, the main object must have been prediction. The list of numbers is interrupted by nine pictures portraying frightful deities and omens. They apparently illustrate the dire consequences attending the onset of each danger period.

How did the predictions work? The table seems to suggest that the priests realized their predictions would be satisfactory if they used sums of only two numbers, 148 and 177. After accumulating sufficient observations over several decades, the priests probably discovered that a lunar eclipse would never happen without a combination of these two numbers. Their rule for prediction may have been as follows: once a lunar eclipse happens, then the next will follow on one of three possible dates in the near future: (a) 177 days later, (b) 325 (177 + 148) or (c) 354 (177 + 177). Assuming that a simple rule of this kind was devised, the priests would find that their forecasts were correct a little more than fifty percent of the time.

One of the leading experts on Maya astronomy, Anthony Aveni, considers the implications of the Dresden eclipse table and concludes:

> The reduction of a complex cosmic cycle to a pair of numbers was a feat equivalent to those of Newton or Einstein and for its time must have presented a great triumph over the forces of nature.[71]

Moreover, not content with this discovery, the composers of the eclipse table required that their predictive technique should harmonize with other important cycles, above all the Sacred Round of 260 days.

They seem to have achieved this by the careful selection of an overall length for the table; it covers just under thirty-three years. The total number of days corresponds closely to forty-six turns of the sacred calendar, and is also a good match for 405 lunar

Lunar eclipse table from the Dresden Codex.

months.[*] In other words, when the priest found himself at the end of the table, he would be able to go straight back to the beginning all over again without losing his place either in the sequence of calendar days or in the phases of the moon.

As presented in the pages of the Dresden book, this period of 405 moons allows us to arrive at an average for the length of one lunar month that falls only seven minutes short of the accepted modern value. Furthermore, according to one recent theory, another set of figures may represent a correction scheme allowing even greater accuracy. If the theory is right and the corrections were made, then the precision of the eclipse table is on the order of *one day in about 4,500 years*.

We can only marvel at the skill with which the Maya strove to combine significant number cycles with such exact standards of skywatching. A long series of observations of lunar phases, averaged over the years, as well as of actual eclipses themselves, must lie behind the whole system.

While it resembles the numerical methods of the Babylonians, this approach is utterly different from the interest in orbital motions first developed by the Greeks, which is still so fundamental to our own thinking. It also forms a contrast to the apparent concern of megalithic astronomers with the orbital swings of the sun and moon along the horizon. Significantly, no convincing alignments to moonrises or moonsets have been detected at a Mayan site, supporting the conclusion from the bark books that it was all done simply by counting the phases of the moon.

Above and opposite, Venus almanacs from the Dresden Codex.

THE SUPER-NUMBER OF VENUS

A further five pages of the Dresden book are connected with detailed predictions of the planet Venus, specifically its cycles of appearance and disappearance as the Morning and Evening Star. In the middle of each Venus page is a drawing that depicts the Morning Star as a fierce armed god spearing a victim. The planet's first appearance in the sky was evidently no less dreaded than the onset of an eclipse.

The basic design and layout of the Venus tables are quite similar to those of the eclipse pages; there was the same concern to achieve extraordinary precision while preserving harmony with the all-important sacred cycles and numbers.

A further demand met by the compilers of the table was that Venus events should be celebrated on particular auspicious days. An earlier generation of scholars assumed that there were puzzling errors in the table, but these have now been shown to relate actual appearances and disappearances of the planet to the nearest appropriate day in the calendar. In fact, the priests were once again seeking a compromise between the real movements

---

[*]To be precise, there are 11,958 days in the table, close to the 11,960 days of 405 lunar months and 46 Sacred Rounds ($46 \times 260 = 11,960$).

observed in the sky and the requirements of religious beliefs and customs. They were doubtless using the table to work out the correct days for staging particular Venus rites and invocations.

In short, the Venus pages bear little resemblance to a modern astronomical table. They were *not* intended to provide a precise, short-term reckoning of when the planet could be next expected to appear or disappear in the sky. No attention was paid to the problem of anticipating immediate fluctuations in the planet's timetable. These variations mean that one Venus year can consist of 587 days, while the next has only 581, but it was the overall average of 584 and the significance of this number which deeply preoccupied the Mayan priests. Their chief concern was to ensure that *in the long term* the Venus observations would remain perfectly in step with several other calendar counts. As in the case of the lunar tables, they achieved this result by devising a phenomenally accurate correction scheme, so that the tables would function perfectly for century after century.

Besides this curious blend of long-term exactness and short-term inaccuracy, perhaps the most interesting feature of the Venus pages is a pair of dates associated with the correction scheme. The purpose of these dates seems to have been to relate the starting point of the table in historical times to the birth of Venus, a legendary event that occurred just before the beginning of the present cosmological era. This exercise was almost identical to the case of the Temple of the Cross described earlier. There the inscriptions related the actual dedication of the temple to the imaginary accession of an impossibly old ancestral mother in the previous epoch. In both instances, the object seems to have been the same: to link the two events in a harmonious way by deliberately inventing a time interval between them that would possess extraordinary astronomical significance.

Thus the mythological birth of Venus was selected so that it fell 1,366,560 days before the starting date of the table. This number lent sanctity to the Dresden book because it can be divided exactly by *at least seven* separate calendrical or planetary cycles. Indeed, its properties are so remarkable that it has been called "the super-number of the Dresden Codex."

Once again we see how Mayan astronomers sought a sacred framework for every activity in which they were engaged. This was their compelling drive, even if it involved such blatant

The "super-number" of the Dresden Codex:

1,366,560 =	
260 x 5,526	(Sacred Round)
365 x 3,744	(Solar year)
584 x 2,340	(Venus interval)
780 x 1,752	(Triple Sacred Round or Mars period)
2,920 x 468	(5 x Mars period)
18,980 x 72	(Calendar Round—reconciles Sacred Round and Solar Year)
37,960 x 36	(Double Calendar Round)

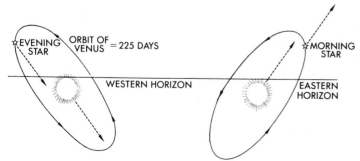

ORBIT OF VENUS = 225 DAYS

Venus always appears close to the sun. These two diagrams show Venus at its brightest, appearing just after sunset as the Evening Star, and just before dawn as the Morning Star. Modern astronomers know that Venus takes 225 days to revolve around the sun. For the Maya, however, the important figure was 584 days, the interval between identical appearances of the planet as seen from the moving earth.

contrivances as "super-numbers," or the timing of Venus ceremonies on different days from the actual appearances and disappearances of the planet.

TOWERS OF THE VENUS GOD

While the final shape of the Venus table owes much to artificial devices of this kind, it must nevertheless have been based on many years of genuine, meticulous observations. One of the most exciting discoveries of the 1970s was that alignments to Venus were incorporated in the layout and architecture of important Mayan buildings, such as those already described at Uxmal.

Of all these buildings, none is more intriguing than the Caracol of Chichén Itzá. It may be the very building from which the observations necessary for the Dresden Codex were undertaken. The document contains evidence that it was based on earlier records compiled around 1000 A.D. in the northern Yucatán. The upper structure of the Caracol dates to about the same era, and so there is a definite possibility that the priests of Chichén Itzá were responsible for developing the Venus scheme preserved in the Codex.

In its original state, when the plaster-covered walls and decorative rain-god masks were freshly painted, the squat, cylindrical tower of the Caracol must have possessed a certain gaudy magnificence. Today the ruins exhibit several peculiarities—notably, a lack of symmetry about the arrangement of nearly all the main walls, doors, and stairways. The upper platform on which the Caracol stands is of irregular polygonal shape, and is not set parallel to the frontage of the main lower platform beneath. But a little ledge near the top of the upper stairway (supporting two small columns) *does* seem to be parallel to the lower frontage. Even stranger, the front door does not face the front. Once inside the Caracol, the visitor encounters a second circular wall with four inner doorways positioned asymmetrically in relation to the outer ones.

The Caracol of Chichén Itzá, a Mayan observatory in the Yucatán, as it must have appeared around 1000 A.D. and in its partly restored state today, opposite.

According to Anthony Aveni, Sharon Gibbs, and Horst Hartung, who published a detailed study of the monument in 1975,
the reason for all these irregularities lies in the numerous astronomical alignments present in the structure. In fact, Aveni and
his colleagues were able to identify twenty out of a total of
twenty-nine major directions incorporated in the architecture.
The result is impressive when we consider that most of them are
alignments to the sun or Venus which would have been directly
useful in the calculations of the Mayan priests.

The most prominent orientation is the main façade of the lower
platform, which faces the direction of the last gleam of the sun at
the summer solstice with an error of less than 2°. However, about
the same error, or slightly less, is involved if we prefer another
target, the maximum northern swing of Venus setting on the
horizon.

This orientation seems to have been deliberate, because it is
duplicated in the positioning of the little ledge on the stairway
above, set conspicuously out of line with the surrounding stairs.
The likelihood that Venus was the target is confirmed by the
colors with which the ledge and the small twin columns standing
on it were painted. Traces of black paint remain on the northern
column and the ledge beneath it, while the southern was red.
Several versions of the myth of Quetzalcoatl, the Venus god, are
preserved in native manuscripts compiled shortly after the conquest. They refer to the location of the divine hero's death as "the
place of the black and the red." The symbolic pairing of the

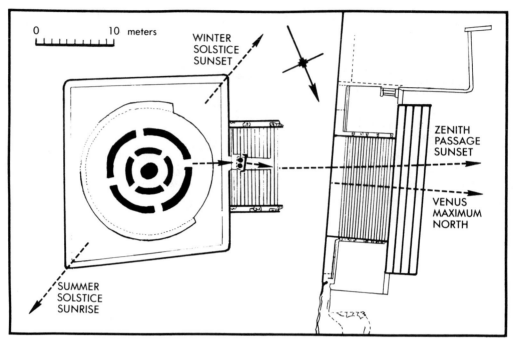

0 10 meters

WINTER SOLSTICE SUNSET

ZENITH PASSAGE SUNSET

VENUS MAXIMUM NORTH

SUMMER SOLSTICE SUNRISE

Plan of the Caracol, above, *indicates major astronomical alignments partly accounting for its odd layout. Plan of the three surviving windows of the Caracol with their probable alignments to Venus and the sun,* below. *Astronomical targets indicated by the southern window,* opposite, *remain uncertain.*

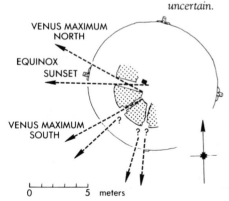

VENUS MAXIMUM NORTH

EQUINOX SUNSET

VENUS MAXIMUM SOUTH

0 5 meters

colors probably refers to the direction west, where the sun and Evening Star set, and where night and day—the black and the red—merge. The arrangement of the black and red columns on the upper stairway of the Caracol strongly suggests that the priests were preoccupied with the setting of the Evening Star as they watched the horizon in this direction.

The most interesting part of the observatory, however, is the upper story of the tower, where a small rectangular viewing chamber was positioned. A circular passage led from the inside up to this observing room; it was the spiral form of this passage that inspired the building's Spanish nickname, Caracol, meaning "snail." The whole of this upper story fell into ruin toward the end of the last century and has been subject to repairs since then. Originally, it seems eight narrow windows faced outwards from the viewing chamber, of which only three survive.

The detailed measurements carried out in the 1970s led to new conclusions about the astronomical properties of the windows, disposing of an earlier theory that alignments to the moon were involved. The Mayan observer seems to have lined up the inside and outside edges of the viewing slots; a pair of these diagonal directions correspond to the extreme northern and southern setting limits of the Evening Star along the horizon. It is natural to assume that the vanished windows on the ruined east side of the tower were lined up in a similar fashion with the dreaded rising of the Morning Star. The thorough investigations of Aveni, Gibbs, and Hartung scarcely leave room for doubt that the Caracol was not only a temple sacred to Quetzalcoatl, but also a kind of instrument for watching the critical cycles involved in the motion of Venus.

It was a highly unusual building, but apparently not unique. Until recently, two other towers in the northern Yucatán exhibited features closely modeled after the Caracol, although both are now reduced to shapeless heaps of rubble. The first is located near Mérida, the modern capital of Yucatán; this is the huge walled city of Mayapán, which rose to prominence after Chichén Itzá was deserted in the thirteenth century A.D. Today the site is completely overgrown, but one can still scramble up the largest of its grassy mounds, clearly a second-rate imitation of the great Castillo pyramid which so impresses visitors to Chichén Itzá.

Close to the pyramid of Mayapán are the remains of its tower, now almost a total wreck. In 1980 it was difficult to make out the base of its circular wall among a tangle of tree roots. Our knowledge of the building comes mainly from two separate accounts and sketches prepared shortly before it was severely damaged by lightning in 1867. These records make it clear that the frontage quite closely resembled that of the Caracol at Chichén Itzá, and may actually have had an upper story with windows for observations. Apparently, there was only one door, giving access to a circular passage with a solid core of masonry in the middle, again reminiscent of the Caracol. Today only a short stretch of the lower rectangular platform can be measured, and this gives no real clue about the observations that may once have been carried out inside the tower which stood on this base.

The other circular building has a wilder and more romantic

setting, at the edge of a remote coral beach at Paalmul on the east coast of the Yucatán, opposite the island of Cozumel. When it was first described in the 1920s, the building was compared to the turrets of a battleship. This image was suggested by its unusual configuration, consisting of an upper tower with twin concentric circular walls positioned on two substantial lower oval-shaped platforms. In each part of the upper tower there was a rectangular room, again reminding us of the viewing point high up in the Caracol.

The expectations of the visitor to Paalmul, aroused by the fringe of coconut palms and the delicate coral pebbles along the shore, are dashed by the ocean-ravaged cone of rubble which is all that remains of the building. Even the solitary back wall of one of the upper rooms, the only intact structure measured by Aveni and Hartung in 1976, was no longer standing when I visited the site in 1980. The orientation measured in 1976 suggested that the tower faced away from the sea in a northwesterly direction. The Mayapán tower probably also faced northwest, as does the Caracol today.

All three towers are located at approximately the same latitude, even though they are separated by many miles. This could be a coincidence, or it could mean that they were deliberately planned together to provide independent confirmation of the movements of Venus.

With the help of alignments such as the ones incorporated in the base and windows of the Caracol, the priests must have patiently arrived at their superbly accurate long-term averages for the planet. As well as providing the essential basis for the Venus almanacs, the alignments probably also enabled the priests to make the short-term predictions not provided by the tables. The position of the planet in its eight-year swing along the horizon is related to the length of the intervals of appearance and disappearance, so that estimates of when the Morning Star was next due to rise were, in fact, possible. Mayan knowledge of the movements of Venus must have been thorough and systematic.

Did their interests extend to the other planets? At the present stage of research, it is difficult to be sure. Three pages of the Dresden Codex are filled with dates and numbers that may relate to the orbit of Mars. Unfortunately, this period of 780 days happens to be exactly equivalent to three turns of the Sacred Round ($3 \times 260 = 780$). If this fact was appreciated by the Maya, it would have been an extra incentive for watching the red planet. But the coincidence also prevents us from distinguishing between this possible Mars table and the many other 260-day almanacs featured in the Codex.

The full reach of Mayan astronomical skills remains uncharted. The question of whether they recognized the orbital periods of the other main planets—Mercury, Jupiter, and Saturn—is currently the subject of lively debate as well as other possibilities such as the existence of a Mayan zodiac. The patronizing view of an earlier generation of investigators—the idea that religion blinded the Maya to a true understanding of the sky—has largely vanished. In its place, scholars of many disciplines now accept the spiritual impulses of the Maya as a vital driving force, and try to absorb themselves in this unfamiliar outlook. Only then is it possible to appreciate the skill with which religious dogma was united with precise observation.

To the long-established objects of Mayan studies, the hieroglyphs and the bark books, scholars may now add evidence from such neglected sources as the orientations of temples and plazas. These advances in understanding all contribute to a more solid and complex picture of the Mayan mentality. It challenges some common assumptions about our own scientific heritage, notably, that a cultural and technological background similar to our own is essential to the development of intellectual qualities we value highly. Accuracy, precision, and the capacity to make sustained, logical observations of natural phenomena—all played a part in the exotic Mayan world of sacred numbers and supernatural ancestors.

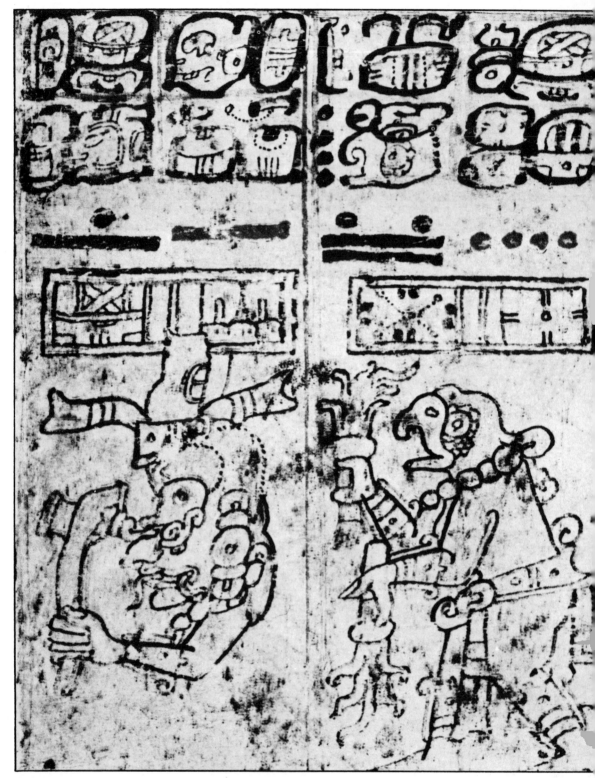

Detail of Mayan numerals and glyphs from the Dresden Codex.

PART V

Conclusion

17. From the Maya to the Megaliths

WERE THERE "WISE MEN" IN WESSEX?

Skywatching in the New World encompassed a vast range of skills and abilities, from the informal lunar calendars of many North American hunting tribes to the startling precision of the Mayan bark books. How can such a diversity of traditions help to answer the general question of how and why ancient astronomy was carried out? For instance, it would be rewarding if we could take a case history from the Native American record and use it to fill in some of the "missing gaps" where our knowledge of astronomical practices is only fragmentary, as in the case of the European megaliths. But is it possible to draw lessons from the nineteenth- and twentieth-century anthropologists and then expect these lessons to illuminate the remote past of another continent?

If we read the anthropologists' accounts of Native American tribes, we occasionally come across remarkably close correspondences with the astronomical theories advanced to account for prehistoric standing stones in Europe. For example, the Thoms suggest that megalithic skywatchers planted wooden stakes to mark their positions on the ground on successive days or nights. The stakes would provide an accurate measure of the changes in the risings and settings of the sun and moon.

Turning to North American sources, we find that the nineteenth-century Nootka of the Northwest Coast were concerned with establishing a period of four to five days "where the sun sits down" on the horizon at the time of the solstices. This period was defined by two observers who lined up a pair of sticks with the sun's disc each day. The Nootka were keenly interested in predicting the date of the winter solstice because their private nocturnal rituals, vital for securing success in the whale hunt, had to be timed carefully. The prayers and anointments were to be performed as the moon was waxing and the days were growing longer, or bad luck was sure to follow. Superstition, rather than science, was the motivation, but at least this example proves that an observing technique similar to those proposed by the Thoms was indeed carried out by Native American astronomers.

But this type of comparision has its limitations. If we probe more deeply in the hope that the Nootka will enlarge our understanding of the megalith builders, the result is disappointing. In fact, a detailed comparison of the Nootka with the Neolithic Britons seems thoroughly inappropriate. The Nootka were celebrated for their adventurous whale- and seal-hunting exploits, although their subsistence more often depended on the vast salmon runs of the Northwest and on other inland game and vegetable resources. Access to many of these resources was an inherited privilege controlled by elite families. In common with most other tribes of the Northwest Coast, the Nootka were organized in a strongly hierarchical fashion, with a class of permanent slaves at the bottom. A "shell-money" economy was in operation, which meant that the Nootka were virtually the "bankers" of the Northwest; the tusk-shaped *dentalium* shells used for currency were most commonly found in waters off their beaches. In fact, the extraordinary abundance of their surroundings must have encouraged the richness of their social and ceremonial life.

Was the social world of the stone circle builders as complex as this? No artifacts hint at the existence of a Neolithic currency, or of such a rigid, elaborate class system as was present on the Northwest Coast. Many of the megalithic observing sites studied by the Thoms are located in a coastal environment in western Scotland, but any attempt to compare the two regions in detail would be far-fetched.

If we seek meaningful comparisons between the Old and New Worlds, then it is not enough to pick out superficial coincidences between two cultures. To draw a more illuminating parallel, it would be better to choose a Native American culture with a social background roughly similar to that of Neolithic Britain.

But the task of deciding exactly what kind of social order was present in Britain over four thousand years ago is not easy. As we have seen, the investigations of standing stones and stone circles frequently yield only scanty and problematical evidence. On the other hand, archaeological discoveries of ordinary domestic dwellings are rare and offer even less scope for speculation. Perhaps the most useful information of all comes from several major excavations in southern England which took place in the late 1960s.

These excavations, conducted by Geoffrey Wainwright for the Department of Environment, disclosed a remarkable series of large circular wooden buildings, marked by rings of postholes which had originally been dug down into the chalk bedrock. This distinctive style of circular architecture emerged in southern England by about 2500 B.C. At three sites—Marden, Mount Pleasant, and Durrington Walls—circular structures were located

Northwest Coast chief proudly displays a copper, a prestige object used in competitive displays of wealth during the last century.

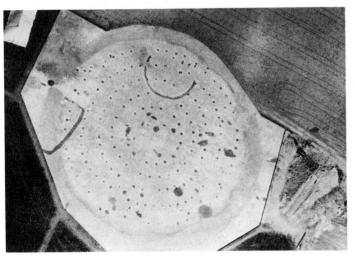

Air view of the circular wooden building at Mount Pleasant, Dorset, England, excavated by Geoffrey Wainwright in 1970. The postholes of this 130-foot diameter building show up as dark patches against the chalk bedrock.

inside irregular formations of banks and ditches; these earthworks represented impressive efforts of manpower and engineering. They were the largest of the so-called henges, of which the best-known examples are Stonehenge and Avebury. Actually, archaeologists loosely use the term "henge" to apply to a wide variety of earth enclosures that may well have served different functions.

What, then, was the purpose of the circular buildings erected inside certain henges? Were these public structures set aside for the ceremonies of a community, or were they the private residences of a chief and his retinue?

The objects discovered in Wainwright's excavations do not, in fact, suggest a clear-cut distinction between ceremonial and domestic activities. At Durrington Walls, about two miles from Stonehenge, great quantities of ordinary refuse, including much smashed pottery, numerous arrowheads, flints, and antler tools, were unearthed inside a vast enclosure over a quarter of a mile across. These finds were particularly dense near one of the entrances to the earthwork, and also inside and around the two circular buildings identified at the site. There is little reason to doubt that everyday domestic activities could account for much of this debris.

On the other hand, the imposing layouts of the largest buildings at Durrington and Mount Pleasant suggest a measure of formality appropriate to public or ritual functions. The mere effort of gathering wood for the major building at Durrington—over 260

View of the excavated ditch near the entrance of the Marden henge monument, Wiltshire. As visitors entered or left around 2000 B.C., they discarded objects, such as animal bones and antler tools, in the ditch.

tons of oak timber, including twin entrance posts that probably weighed five tons each—implies a highly coordinated, purposeful plan. Moreover, in the case of one building at Mount Pleasant and another known as the Sanctuary, near Avebury, the structures were eventually replaced by megalithic stones. These were set out so that they respected the old wooden foundations, as if to commemorate them as places of sacred significance. The ritual purpose of the Sanctuary building is also indicated by the large quantity of human bones unearthed there during the eighteenth century; its original function seems likely to have been a charnel house.

The scarcity of burials at the three great henges investigated by Wainwright indicates that these sites served a different end. To quote the excavator himself, it appears that "in earthworks of the Durrington–Marden–Mount Pleasant class we are seeing secular centers dominated by large public buildings which were accompanied, as at Durrington and Marden, by less massive structures for which a purely domestic use might be deduced."[72] The tradition of circular wooden architecture seems to have been adapted for more than one purpose, a conclusion that almost certainly applies to stone circles as well.

It seems unlikely that the circular wooden buildings incorporated any elaborate astronomical or geometrical knowledge. Take the case of Woodhenge, a structure located just outside the southern bank of Durrington Walls. Originally, Woodhenge was supported by six concentric rings of wooden posts with an overall diameter of more than 150 feet. In an exhaustive study of the layout of these posts published in 1981, the mathematician A. H. A. Hogg concluded that the posts were placed so irregularly that it is impossible to reconstruct any original design that may have been intended by the builders.

While the architects must surely have worked with numbers and measurements in planning such substantial structures, it does not necessarily follow that they were astronomer-priests. Indeed, the nature of the authority exercised at Woodhenge and Durrington is difficult to identify. It may be wrong to assume that an elaborate hierarchy of chiefs, retainers, nobles, and priests was necessary to organize the construction and maintenance of these centers.

Fresh estimates of the manpower involved in raising the henges have drastically reduced the calculations originally made in the 1960s. The earthworks at Durrington are now thought to have consumed half a million man-hours and the largest building about eleven thousand. These figures seem impressively high, until they are compared with estimates of the local prehistoric population (possibly as much as fifty thousand in the Wessex region). The calculations are largely guesswork, but it is clear that building a henge would not put a serious strain on the ordinary subsistence routine, particularly if the project were spread over more than a single season.

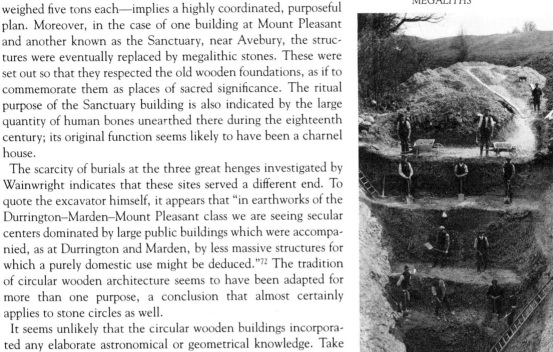

The vast size of the henge at Avebury, Wiltshire, is conveyed by this photograph taken around the turn of the century during the excavations of H. St. George Gray. Neolithic laborers dug the 30-foot ditch into solid chalk solely with antler picks and bone shovels. Did such an operation put a strain on the resources of Neolithic society?

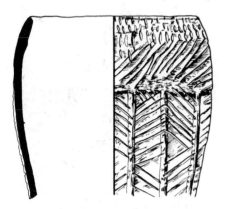

A typical example of Grooved Ware, excavated from Durrington Walls, Wiltshire.

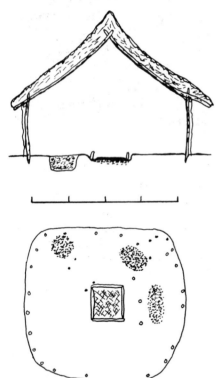

A simple domestic dwelling associated with Grooved Ware, excavated at Trelystan near Welshpool, Wales, and dating to about 2500 B.C. Scale in meters.

While the skill and forethought of the planners should not be underestimated, it is probably a mistake to imagine them as ruthless aristocrats driving a submissive horde of peasant laborers. The monuments erected in Britain before about 2000 B.C. are more likely to represent expressions of communal values and beliefs; they were not costly celebrations of powerful individuals.

The social picture emerging from the digs at the huge henges is obviously full of uncertainty. But Euan MacKie has proposed a bold theory about exactly who the inhabitants of Durrington Walls, Marden, and Mount Pleasant were. In a book with the intriguing title *Science and Society in Prehistoric Britain*, published in 1977, and in later writings as well, MacKie suggests that these sites were "permanently inhabited by a specialist group of people, probably members of religious orders, who were supported by the rest of the agricultural population."[73]

Astronomers and mathematicians were present among this privileged elite. Their position afforded them the leisure to develop and memorize the elaborate skills essential to Thom's theories. In fact, MacKie writes, the great henges of southern England served as a kind of college, "the living and training places of learned orders of astronomer priests and wise men, the activities of which are everywhere to be seen in the standing stones and stone circles."[74] This professional class was supplied with tributes of food and labor by a scattered population of rural peasants.

If we are to take this seriously, we would expect such a specific explanation of the henges to be matched by equally specific archaeological evidence in support of it. But the interpretation of most of the material unearthed at the henges is highly ambiguous. For instance, MacKie supposes that the astronomer priests drank from a special "upper class" pottery known to archaeologists as Grooved Ware. This flat-bottomed style of vessel is associated with henges throughout England and Scotland, and was obviously widely traded by the builders of these sites. Yet Grooved Ware also found a use in ordinary domestic settings, for fragments appear at the remains of humble rural homesteads. Besides, most Grooved Ware pots were very poorly fired in comparison with other types of the same period, and their roughness was surely unfitting for a privileged class.

Because the archaeological picture is so incomplete, MacKie turns to the Maya. In his book, he wonders if the standing stone sites could be "the architecturally cruder—yet ceremonially just as complex—equivalents of the Maya temples. . . ."[75] If it is fair to make comparisons between the two cultures, MacKie believes that it is also reasonable to expect the stone circle builders to have had an elite class of priests who, like the Maya, were skilled in the prediction of eclipses and planetary motions.

Is MacKie right? Does the indisputable evidence of highly refined observations among the Maya make similar claims for ancient Britain plausible?

His picture of the great henges as "ceremonial centers" supported by farmers dispersed throughout the surrounding countryside was inspired chiefly by the example of the Maya. According to MacKie, the comparison is a good one, since both the Neolithic Britons and the later Maya were nonurban, stone-using societies. The privileged Maya, who included skilled astronomers and mathematicians, lived, according to MacKie, "in special ceremonial centers most of which were isolated groups of stone buildings among the scattered dwellings of the rural population, relatively small amd composed solely of temples, priestly residences and 'palaces.' "[76] Unfortunately, this represents an outdated view of how the Maya lived. The advanced intellectual achievements revealed by the Dresden tables and the Caracol of Chichén Itzá were the product of a very different kind of setting than this description suggests.

REVEALING THE MAYA WORLD

MacKie did not deliberately mislead his readers, for the idea of the "ceremonial center" was for many years the most widely accepted explanation of Mayan sites. The term was used as early as 1931 by perhaps the most distinguished of all Mayan experts, the late J. Eric S. Thompson. Besides accomplishing important advances in the reading of hieroglyphs, Thompson also awoke public interest with such lively books as *The Rise and Fall of Maya Civilization* (first published in 1954).

In his popular accounts, Thompson often repeated his concept of the "ceremonial center." It meant that Mayan pyramids and palaces were the residences of the ruling and professional classes, while the rest of the population lived elsewhere. Back in 1932, Thompson explained that "the term 'city' as applied to the Maya ruins is a misnomer. They were almost certainly religious centers to which the people, who lived in small settlements scattered over the surrounding country, repaired for religious purposes and, possibly, to attend markets and courts of justice."[77] Thompson frequently emphasized the class divisions of the ancient Maya in his popular writings. With little evidence to support his view, he supposed that the collapse of their civilization was brought about by a peasant revolt. It was his suspicion that "in city after city the ruling class was driven out or, more probably, massacred by the dependent peasants, and power then passed to peasant leaders and small-town witch doctors."[78]

The causes of the Mayan decline in the southern lowland region remain a mystery today, but they were certainly more complex than a single revolutionary uprising. The process of disintegration took at least a century, beginning around 800 A.D., and its impact was felt sooner in some regions than in others. The construction of new buildings and the carving of calendar inscriptions stopped, while the settlements themselves eventually lost much of their population. There is little evidence of the violence implied by Thompson's views. One can only assume that a

A typical Mayan dwelling in the Yucatán. The plan of these domestic structures has changed little since the days of the ancient Maya.

View from the top of Temple I over the magnificent Great Plaza of Tikal, with pyramids dating mainly from 700–900 A.D.

complicated network of political alliances gradually broke down under a number of stresses; such factors as competition for trade or the spread of diseases could have played a part. Meanwhile, the northern region of the Yucatán seems to have escaped the progressive collapse that overtook the lowland sites. The Mayan civilization not only survived there, but was also enriched by the building of many great settlements such as Uxmal and Chichén Itzá, all belonging to the period after 800 A.D.

As mentioned in the last chapter, Friar Landa informs us that during the sixteenth century astronomer-priests still commanded influence in the Yucatán. There is no way of assessing their importance during the earlier Classic age of pyramid and palace architecture, except by noting the prominence of calendar inscriptions and such structures as the Caracol. But there is little doubt that the priests were *not* the solitary residents of the Mayan building complexes, as Thompson believed. Recent dramatic archaeological discoveries cast serious doubt on his notion of the "ceremonial center."

During the 1970s and 1980s, mapping work in and around the Mayan sites has intensified. Archaeologists have recorded not only conspicuous temples, but also barely visible domestic structures such as house mounds. These low rectangles of earth provided a foundation for simple wooden, thatched huts, probably identical to those still built by the modern Maya throughout the Yucatán today.

The effort of plotting these humble structures has proved rewarding. At many sites the house mounds are clustered around

the central stone buildings in surprising numbers. At Dzibal-
chaltún, a settlement in the Yucatán located close to Mérida, the
population must have been no less than two hundred thousand,
nearly two-thirds of the total of modern Mérida itself. While this
was clearly exceptional for the region, the map of Tikal, the great
lowland center in Guatemala, compiled by a team from the
University of Pennsylvania during the 1960s, tells a similar story.
The "downtown" core of Tikal, comprising 2½ square miles of
pyramids and palaces, was surrounded by some 75 square miles of
outlying "suburbs," consisting mostly of earthen house mounds.
An estimated seventy-two thousand people lived in the city at its
height during the eighth and ninth centuries A.D.

Indeed, "city" is *not* a misnomer for many Mayan sites, even
those built on a smaller scale than Tikal. While the Mayan cities
never reached the stage of highly organized planning apparent at
Teotihuacán in central Mexico, neither were they "ceremonial
centers" corresponding to Thompson's vision.

Side by side with these urban revelations came an equally
profound change in the understanding of Mayan subsistence
methods. The high density of the city populations could scarcely
have been supported by the rudimentary "slash and burn" agriculture traditionally practiced in the tropics. Without intensive
methods to increase the yield of their forest clearances, maize
growers inevitably remain dispersed throughout the landscape. By
contrast, the density of settlement around Tikal was probably
eight times greater than conventional Mayan methods can support today.

How had their ancestors managed to overcome the limitations
of tropical farming? In the late 1960s, the first evidence of
irrigation canals and terraced fields came to light on the western
and northern fringes of the Mayan lowlands. Thousands of square
miles of artificially modified terrain could be recognized from the
air. The climax came in 1979, when a U.S. Space Agency
aircraft was chartered to fly over wide areas of Belize and

*A section of the map of Tikal, Guatemala,
compiled by archaeologists from the
University of Pennsylvania.*

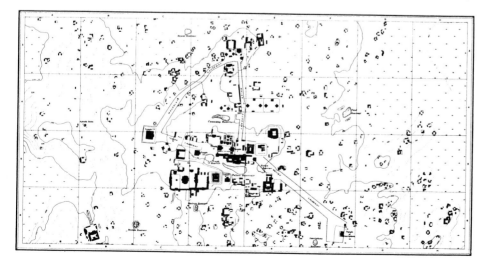

Portrait of a Mayan ruler from Seibal in the Petén jungle. During the little-understood events of the southern Mayan collapse, Seibal appears to have been taken over by non-Mayan outsiders, before it was finally abandoned around 930 A.D.

Guatemala. The aircraft carried a radar scanning device originally designed for monitoring the surface of Venus. In the resulting radar imagery, it was possible to trace latticelike patterns of lines running across swamps and the edges of lagoons and rivers. When checked by ground survey, the lattices proved to be canal systems reaching far out into the jungle around the major Mayan sites. The farmers had scooped up mud and vegetation from the canal bottom and piled it up in rectangular ridges or plots. These raised fields were phenomenally productive, with a potential yield over ten times that of typical forest clearance. The *minimum* population estimate for the total area of canals identified by the Venus probe is no less than 1¼ million people. How different their way of life must have been from the scattered peasantry envisaged by Thompson!

The final ingredient in the drastic appraisal of the Mayan way of life was contributed by advances in the understanding of hieroglyphic writing achieved during the 1970s. An example of the depth of this new knowledge was given in describing the ancestors and descendants of Pacal, ruler of Palenque. As we have seen, Mayan mathematical and astronomical skills were not dedicated to an abstract pursuit of learning for its own sake, as the Thoms believe their megalithic observers to have been. Instead, Mayan observations played a crucial role in supporting the Mayan ruler and his family, who were caught up in complex alliances and rivalries with the dynasties of other cities. Scholars of the hieroglyphs can now appreciate something of the intricacies of Mayan political thought; the extraordinary preoccupation with mythological ancestor figures like Lady-Beast-with-the-Upturned-Snout is a good example. The Mayan astronomers' pursuit of time cycles was not an isolated, obsessive drive, but formed part of a rich background of values and beliefs upholding the sacred authority of the ruling family.

Of course, their knowledge of writing distinguishes the Maya from many of the other cultures reviewed in this book. As Euan MacKie is the first to admit, the literacy of the ancient Mayan skywatchers is a major stumbling block in any attempt to compare them with the Neolithic Britons. The highly precise predictions of the Maya were the product of centuries of written records, and represent a different order of achievement from the solar or lunar alignments incorporated in such monuments as Newgrange, Stonehenge, or the Clava cairns. Though the great henges at Durrington Walls and Mount Pleasant must have been remarkable places in their heyday, neither the density of settlement nor the social complexity involved in these sites can have approached that of the Mayan cities.

CHIEFS, PRIESTS, AND CALENDARS

Since the Mayan comparison with prehistoric Britain is clearly inappropriate, what about the preliterate societies of North America? Does a study of the Pawnee or the Chumash lead to conclusions that apply to all peoples without writing, in the ancient past as well as in recent history?

Perhaps the most striking aspect of the North American record is the great variety of social settings in which elaborate astronomical skills and ideas flourished. The Hopi sunwatchers, whose performance might be casually criticized by anyone, were in a very different position from the Chumash astrologer or *'al-chuklash*, who belonged to the *'antap*, an esoteric, elite cult organization with widespread political influence. Despite this contrast in their social backgrounds, the sky lore of the Hopi priests was no less elaborate than that of the Chumash. The more formal and institutionalized priesthoods such as the *'antap* engaged in precise solar observations, but so, too, did the *Pekwin* of Zuni and the Mayan calendar priests of Nebaj. There seems to be no simple relationship between the degree of skills and knowledge cultivated by the preliterate skywatchers and the kind of society to which they belonged.

Stephen McCluskey has calculated that a long-distance solstice alignment accurate to one-third of one degree was the basis of a series of calendar observations undertaken by Hopi sun priests about a century ago. The special conditions of excellent visibility in their stark desert landscape may have encouraged such a

Engineering skills comparable to those of the European henge builders are evident at sites of the Hopewell culture in the American midwest (c. 100 B.C.-300 A.D.). Studies of the astronomy and geometry of these sites have proved difficult, due to erosion of many mounds. Air view of part of the immense earthwork complex at Newark, Ohio, shows octangonal enclosure, 1500 feet across, now a municipal golf course.

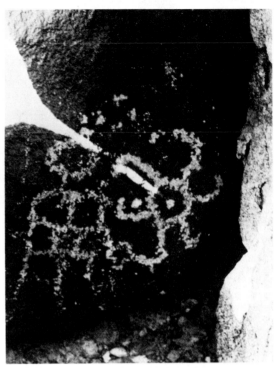

A dramatic development occurred as this book was going to press. Radio astronomer Robert Preston and his artist wife, Ann, described their discovery of nineteen prehistoric rock carving sites in Arizona with lighting effects similar to those of the Fajada Butte "sun dagger." Left, carving in Petrified Forest is pierced by a ray 45 days before and after midwinter. At another Arizona site, right, three pointed images cross the panel at midwinter. Clearly such sun shrines were once common in the Southwest.

remarkable level of precision. Nevertheless, the Hopi example suggests that highly accurate calendar observations were not out of place in a relatively egalitarian society.

The motivation for their precision was not, of course, philosophical or scientific inquiry. The Hopi horizon calendars appear to have increased the efficiency and reliability of their efforts to raise corn, but the chief impulse in the minds of the sun priests was tradition. It had always been done that way; supernatural beings such as Masau had decreed these practices of the Hopi since earliest times. Incidentally, it is interesting to consider that none of the accurate sunwatching activities of the Hopi would leave traces for a future archaeologist to uncover, since no permanent recording devices or monuments were involved.

The problems of megalithic observatories such as Kintraw or the Fairy Stone illustrate how difficult it is to establish that long-distance alignments really did exist at a particular prehistoric site. But even if we assume the theories are correct, the presence of an accurate alignment is no guarantee that a complex hierarchy of priests was in control. While it is certainly likely that a higher authority of some kind was responsible for organizing the layout of Stonehenge or Newgrange, the North American evidence shows that many different kinds of chiefs and priests existed.

For example, the kind of society responsible for the dramatic ruins of Chaco Canyon is still far from certain. As discussed in Chapter 11, it may be wrong to assume that powerful chieftains were essential for the planning and maintenance of the towns.

The layout of Pueblo Bonito, for instance, hints at a dual division of society. This may have been similar in principle to the moiety groups of the more-or-less egalitarian pueblos of the Rio Grande today. At the same time, the ingenious "sun dagger" site on Fajada Butte and the incorporation of cardinal orientations and standard units of length in Pueblo Bonito remind us of the well-developed intellectual activity of the Chaco people. The appearance of astronomical alignments, together with evidence of architectural skills and organized manpower, do not automatically imply a highly regimented society, whether at Woodhenge or Pueblo Bonito.

Equally, the presence of a solar alignment at an ancient site may not necessarily mean that an accurate calendar was in operation. The evidence of many megalithic sites suggests that the alignments were connected with the symbolism of death and rebirth as much as with any practical division of time. Some of the most convincing examples of megalithic astronomy—the passage grave of Newgrange, for example, or the Clava cairns—remind one of this association with funerary customs.

In a similar way, at numerous locations in Arizona and California, striking patterns of light shine across carved or painted panels at the time of the solstices; some of these sites may well represent "sun shrines," rather than accurate observatories for fixing a calendar.

Certain skywatchers, such as the Mursi of Ethiopia, paid close attention to the sun, particularly to the solstices, without feeling a compulsion to establish a rigid calendar. A state of disagreements and inaccuracy in calendrical affairs was in fact the normal situation for many North American tribes during the last century. The examples of the Pawnee and other Plains peoples show that a refined calendar was not vital to the development of intricate beliefs about the sky. Long before the invention of agriculture or well-regulated calendars, celestial myths and seasonal observances figured prominently in prehistoric minds.

THE UNIFYING VISION

The one characteristic shared by all the skywatching peoples discussed in this book was a unifying vision that connected their astronomical skills to so many other aspects of their lives. The order perceived in the heavens provided a model that gave form and meaning to the actions of people on earth. Whether their needs and decisions revolved around the right time to plant corn or the proper place to raise a cairn or pyramid to honor a dead lord, the cosmic order provided them with guidance and justification. This conscious use of the sky to create and reinforce social values was quite different from the attitude of an astronomer today. Of course, we know that the path of modern scientific inquiry is influenced by economic and political forces; indeed, if we look beneath the proverbial objectivity of the scientist we might find his calculations serving such ends as guiding missiles and spy satellites. However, it is the essential and explicit concern of the ancient astronomers with human affairs, as well as with beliefs about the natural order, that distinguishes their thinking from ours today. This is not to suggest they were any less capable of logical reasoning than a modern researcher, or less interested in predicting events. But the motivation of their skywatching skills corresponded to different needs and anxieties from those of our technological world.

It is an illusion to believe that we can use modern calculations and surveying equipment as objective tools to discover "what was scientific" in the achievements of ancient cultures. Any attempt to unravel purely scientific impulses in the minds of the sky-watchers is doomed to failure from the start. To do so, in the words of R. J. C. Atkinson, is "to people the prehistoric past with ourselves in fancy dress."[79] The resulting confusions and difficulties are evident from the controversies that beset the astronomy of standing stones in Europe. Ironically, these debates illustrate how scientists often lose their mask of objectivity and become deeply swayed by the fashionable tastes and interests of the age.

The religious or astrological preoccupations of early astronomers are sometimes dismissed as unimportant, or as obstacles that prevented the observers from making true scientific progress. But in many cases there would have been no ancient observations at all without the unifying religious vision that accompanied them, a vision in which there were no distinctions between the human and the cosmic order.

In the case of the Hopi, their solar horizon calendars served to coordinate their complicated cycles of rituals and dances. The calendars were also, apparently, of direct survival value to them in raising corn under the unpredictable conditions of the Arizona desert. In this instance and many others, there is little sense in trying to draw a line between the practical utility of astronomical observations and the beliefs they served.

The distinction between religion and the "practical" or the "scientific" is particularly meaningless in the case of the Maya. A mystical unifying conception *was* the motivation of their mathematics and astronomy. It was expressed in a holy framework of number cycles, to which nearly all their observations of the sky were related, and sometimes deliberately distorted. The sacred scheme was important enough to determine the layout of tombs and temples such as those at Palenque, and to lie hidden in seemingly decorative façades such as that of the Nunnery at Uxmal. These buildings were not observing devices, but instead symbolized the divine celestial and earthly relationships perceived by the Maya. As Anthony Aveni has written, the building orientations of Central America "all emphasize not scientific precision in the use of astronomical instruments (an applaudable western technological feat) but a total awareness of the natural environment and a desire on the part of the Maya to embody Nature's course in their earthly works."[80]

The all-embracing sky vision of the Maya found its most remarkable expression in the inscriptions upholding the political fortunes of Mayan nobles. In the esoteric calculations that ranged back and forth over thousands of years, frequently extending back into mythological time, the human social order was completely identified with the framework of nature. Mayan "science" was an instrument of social control, not a disinterested study of natural phenomena. Obviously, it was a force of immense power and conservatism, for the ordinary farmer must have believed that agricultural prosperity flowed from the political success and ritual performances of the divine ruler, and not the other way around. This reversal of the modern Western outlook makes it easier to understand why political considerations were as important to the Mayan astronomers as their urge for precision and the factual recording of celestial cycles. Two examples of this attitude are the possible staging of the battle depicted at Bonampak to coincide with a special appearance of Venus, and the apparent tampering with historical dates represented by the inscriptions of the Temple of the Cross.

This manipulation of the astronomical record has an interesting parallel in the annals of ancient China, particularly those of the Former Han Dynasty (202 B.C.–8 A.D.). The Chinese Emperor's rule was divinely sanctioned by a blending of earthly and cosmic forces: "The king has Heaven as his father and Earth as his mother; he is the Son of Heaven. . . ."[81] He was the intermediary between the forces of nature and the lives of his people. The main task of the state astronomers was to detect imbalances in this relationship by watching for portents, such as eclipses, meteors, comets, and other unusual phenomena. This responsibility obviously placed them in a position of immense power in the Han bureaucracy. An examination of the annals shows that the scribes had heavily "edited" them with additions, deletions, and alterations. Certain omens, such as eclipses, were reported on dates which were astronomically impossible, suggesting that the importance of securing a sign overrode the Han astronomers' concern for facts. The Mayan observers were probably not as

Chinese diviners using a board game known as liu po *to determine future events.*

formally organized as the Chinese bureaucrats, but the consequences of their shared belief in the divinity of their rulers were similar.

The political hierarchy, social conventions, and natural events experienced by the Maya were all blended into one highly ordered vision. Remnants of their outlook still linger on in the world of the Chamula, described in Chapter 14, where not only ceremonial activities, but even domestic space, is arranged to conform strictly to the movement of the sun. We may admire such a unified scheme, but the restrictions and rigidity of this attitude are also disturbing to the Western mind. Everything is preordained and "put in its place." The conservatism of the Mayan sky vision explains the tenacious hold of traditional beliefs among the rural communities of Guatemala and Mexico today.

What lessons can be learned from the Maya? Does the special mentality that drove their astronomers mean that their achievements are merely an isolated historical curiosity? Or are they relevant to an understanding of how science developed?

A number of intriguing comparisons can be drawn between Mayan observations and those of the Babylonians, which formed the essential foundation for Greek—and subsequently Western—astronomy. Because of their location in the tropics, the Maya naturally related most celestial events to the horizon and zenith, rather than to the zodiacal system emphasized by the Babylonians. But the approach of astronomers in both civilizations was

The traditional ceremonies of the Maya continue in Guatemala, despite many acts of indiscriminate violence directed against them by the ruling military junta. In 1982 an estimated half million Maya were forced to flee from their long-established communities to refugee camps in Mexico.

surprisingly similar when it came to the problem of predicting lunar eclipses and the appearances and disappearances of Venus.

The high precision achieved by the astronomers of both civilizations was possible because of generally similar background forces. The invention of an accurate system for recording dates was obviously one essential; so, too, was a conservative social order which assured the continuity of observations over a long period of time. Although the evidence that megalithic sites such as Stonehenge were used over many centuries implies a similar persistence of traditions, the absence of writing casts doubt on the ability of prehistoric Europeans to predict eclipses.

The attention of the Babylonian and Mayan skywatchers was confined strictly to the bare task of foretelling a recurrence of the same event. For this reason, they were preoccupied with repeating time cycles, rather than with calculating durations (reminding us of the North American attitudes to the flow of time discussed in Chapter 8). In both cases, their technique consisted of applying ingenious arithmetical analyses to a long series of patiently accumulated observations, which for the Babylonians stretched back to at least 1800 B.C. This practice of searching for regularities among the lunar and planetary numbers was an important foreshadowing of the scientific method.

Yet there is no evidence that either the Babylonian or Mayan astronomers acquired an insight into *why* eclipses and other events in the sky actually happened. No explanations or laws resulted directly from their observations, for no questioning of the divinely ordered scheme was necessary. The element of rational inquiry did not exist.

The Greek philosophers shattered the unity of the ancient sky visions, that unity of outlook shared alike by the Babylonians, the Maya, the Native North Americans, and, doubtless, the megalith builders, too. With the liberating influence of democratic discussion in politics, the notion came about that skepticism and debate might also enter into ideas about the natural world. As the pursuit of philosophical questions acquired a value of its own, the sciences were loosened from their bonds with religion and ideology, and began to gather the momentum of the technological age.

This unchained force of scientific skills and knowledge, our inheritance from the civilization of Mesopotamia and Greece, has now led us to the dangers of the nuclear era. When we consider ancient astronomy, we begin to value the complexity and logic of other schemes of thought besides our own. We realize that our framework of ideas developed from only one system of thought out of many that have passed into obscurity. The perspective of other peoples, sometimes glimpsed across a gap of countless generations, reminds us of the shortcomings of our own outlook, and that there are indeed many others. Such an awareness may be essential for our survival.

Glossary

'Alchuklash Astronomer, astrologer and priest-shaman of the Chumash people of southern California. The *'alchuklash* was a member of the *'antap* cult, and exercised considerable political and spiritual influence over the affairs of the ordinary Chumash.

Adobe Unfired mud brick, hardened by exposure to the sun.

Almanac Table of events of an astronomical or other nature, usually arranged sequentially by date.

Anasazi A Navaho word meaning "Old Ones" or "Ancient Ones." The most brilliant prehistoric culture of the American Southwest, centered on Northern Arizona, New Mexico and southern Utah and Colorado. The culture is chiefly associated with the erection of impressive stone pueblos during the "Classic" (Pueblo III) era, from about the 11th to the 14th centuries A.D.

'Antap The most influential religious cult among the tribes of southern California, with priest-shaman members scattered in most of the towns and villages of the region. At a provincial level, the cult was organized into a council known as "The Twenty," with an inner body, "The Twelve," who were ultimately responsible for maintaining the balance between human affairs and the natural environment.

Antizenith The point directly under the observer, opposite the zenith. Also known as the nadir.

Aztec The final great empire of Central America, lasting from the early fifteenth century to the Spanish conquest of 1519–21.

Cacique A term of Arawak, West Indian origin, widely used by the peoples of Latin America and the American Southwest to refer to a tribal leader or chief.

Caracol Spanish name, meaning "snail," for the circular observatory at Chichén Itzá, Yucatán, suggested by its inner spiral staircase.

Cardinal points The north, south, west, and east points of the compass or astronomical horizon.

Ceque The system of forty-one straight-line pathways of sacred ritual, historical, astronomical and even meteorological significance established under the Inca civilization of Peru, and radiating from the ancient capital, Cuzco.

Codex Any Central American manuscript predating the Spanish conquest, or compiled in the native manner during the early years of Spanish power.

Cuneiform   The writing system of cuneiform or "wedge-shaped" characters, developed in Mesopotamia from the 3rd to the 1st millennia B.C. It was used to express several different languages such as Sumerian, Akkadian, Elamite, Hittite, etc.

Cup-and-ring mark   The tradition of engraving on natural rock outcrops practiced in the Neolithic and Bronze Ages in southwest Ireland, northern England, and much of coastal Scotland. The significance of the basic motif, a cup or depression surrounded by concentric rings, remains obscure, though a connection with burial rites seems likely.

Datura   Also known as Jimson weed. A species of plant with funnel-shaped flowers and prickly pods, widely ingested by Native Americans for its hallucinogenic effects.

Daykeeper   The influential shaman or priest of the Quiché Maya of highland Guatemala, responsible for keeping track of the day names and associated prognostications of the Sacred Round, the 260-day calendar.

Declination   The angular distance of a heavenly body from the celestial equator, measured on the great circle passing through the Pole and the body. The astronomical equivalent of geographical latitude, giving the position of an object in the sky.

Demotic   The cursive form of Egyptian hieroglyphic writing, developed for secular use from about the 7th century B.C. to the 5th century A.D. The central inscription on the Rosetta Stone.

Eclipse   A lunar eclipse is caused when the earth passes between the sun and the moon, so the moon falls in the earth's shadow. A solar eclipse is caused when the moon passes between the sun and the earth, blocking the sun's rays.

Ephemeris   A table showing the positions of heavenly bodies on different dates, usually arranged in a regular sequence.

Equinox   The point on the celestial sphere where the sun crosses the celestial equator. The dates of spring and autumn equinox are usually March 21 and September 20. On these dates the sun rises and sets approximately due east and west.

Gnomon   A vertical marker, used for observations and measurements of the sun's shadow as it varies with the passage of the hours and seasons.

Henge   General term for sacred circular enclosures of the Late Neolithic period in Britain. These structures varied greatly in size and purpose; some were clearly the earthwork equivalents of stone circles.

Hieroglyph   The stylized picture-writing developed in ancient Egypt between about 3000 B.C. and 400 A.D. Also used to refer to the pictogram scripts of other civilizations such as the Maya.

Hopi   A Pueblo people of northern Arizona. The word is also used to refer collectively to the half-dozen villages inhabited by the Hopi.

Huaca   A Quechua (Native Peruvian) term for a sacred site, applied to tombs, springs, caves, ruins, crossroads and many other features of the landscape.

Inca   The final great civilization of the Peruvian Andes. Their far-ranging empire lasted for about a century before the Spanish arrived in 1532.

Katchina   (Various spellings). The most widespread cult organization of the Pueblos, usually composed of all male inhabitants over the ages of six to nine, and in some cases female inhabitants, too. The cult is usually identified with actual or mythical ancestors, and with rain clouds. Its members impersonate the Katchina spirits with striking masks and costumes in the course of ceremonial dances, and use Katchina dolls to instruct children about the cult.

Kerb-cairn   A type of burial monument in southern and western Scotland, characterized by a perimeter of flat slabs surrounding a central stone cairn covering one or more cremation burials. Some kerb-cairns may date as late as 1200–1300 B.C.

Kin   A root-word in modern Maya dialects, with the combined sense of "day," "sun" and "time."

Kiva   The sacred underground chamber of the Pueblos, generally circular but sometimes rectangular in shape, and entered by a hatchway through the roof. The meeting place for Pueblo cult groups and religious societies.

Marques de Chasse   The French name for the rows of abstract notches and marks that are often found on Ice Age bone and antler objects in Europe. Alexander Marshack uses the term "notations" to refer to them.

Maya   The largest group of Native Americans north of the Andes, centered on Guatemala, Belize, and adjoining parts of Mexico, El Salvador and Honduras. The height of ancient Maya civilization lasted from about 200 A.D. until its mysterious collapse in the southern lowland region around 790–900 A.D. In the Yucatán, Maya civilization continued until the Spanish conquest.

Megalith   From the Greek, meaning "large stone." A term used loosely to refer to prehistoric standing stones or stone-built tombs.

Moiety   A division of a community into two groups, usually for special purposes such as ceremonies or games. In many pueblos, an individual adopts the moiety of the father. Membership in a Pueblo moiety does not affect the choice of marriage partners.

Neolithic The New Stone Age, marked by the coming of agriculture, dating in the Near East to at least 6000 B.C. and in Western Europe from roughly 4000 to 2000 B.C. In the New World, some plants were domesticated as early as 6000 B.C., but maize was not a staple in Central America until about 1500 B.C.

Obsidian A natural glass resulting from the rapid cooling of lava. Superior to flint, obsidian was much prized in antiquity for the manufacture of stone tools.

Paha The ritual assistant of the *wot* or chief of the southern Californian tribes. The *paha* directed ceremonial gatherings and was responsible for maintaining the cosmic balance affected by each ritual act.

Patolli The Central American version of pachisi, a game which originated in India. The players throw dice to move counters along the arms of a cross-shaped board.

Pekwin The office of Sun Priest at Zuni Pueblo, New Mexico.

Pueblo A term used loosely to refer to the native people of Arizona and New Mexico, including tribes such as the Hopi and Zuni, and also to the adobe- or stone-built settlements occupied by these people. In the latter sense, "pueblo" is also applied to prehistoric settlement sites in the Southwest such as those of the Anasazi.

Radiocarbon dating One of the most useful dating techniques known to archaeologists. All organic material, such as wood, charcoal, skin, shell and bone, contains small amounts of carbon 14, a radioactive isotope of carbon present in the atmosphere. When the organism dies, the carbon 14 gradually decays into the normal form of carbon (carbon 12) at a known rate. The dating technique consists of measuring the relative proportions of C14 and C12 in a sample; the older the object, the less C14 is present. The method is not precise, since the result is expressed in statistical terms, within a given margin of error. It works well for samples up to about 70,000 years old.

Recumbent stone circle A distinctive form of ritual monument erected in northeast Scotland during the later 3rd millennium B.C., consisting of a ring of stones with a large supine stone placed in the west-southwest/south-southeast sector. Cremations are often associated with cairn-like structures in the center, but the circles may have been intended for ceremonies as well as for burial rites.

Sacred Round The 260-day ritual calendar of the ancient and modern Maya, consisting of a sequence of thirteen numbers paired with twenty names for days. The same combination of numbers and days recurs once every 260 days.

Shalako The principal midwinter ceremony of Zuni Pueblo, New Mexico. The Shalako are awe-inspiring masked beings who are "Messengers of the Gods" sent to bless newly constructed houses in Zuni.

Shaman In its strict sense, a shaman is a Siberian priest or medicine man with access to supernatural sources of healing and knowledge. Shamanism, in which spirits communicate with or take possession of the individual, is a fundamental aspect of the religious tradition of native peoples throughout the New World as well as in Asia.

Solstices The two times during the year when the sun is at its farthest distance to the north or south, around June 21 and December 20.

Stela An upright stone slab, usually ornamented with incriptions or carvings.

Supernova The explosion of a star, emitting ten or even 100 million times more light than our sun. Astronomers expect a supernova to occur within a galaxy every 600 years or so.

Theodolite An instrument mounted with a telescope for measuring horizontal, and sometimes vertical, angles.

Transit American term for a theodolite.

Tropic of Cancer and Capricorn The boundaries of the tropical zone at latitudes $23\frac{1}{2}°$ north and $23\frac{1}{2}°$ south. At noon on the day of summer or winter solstice respectively, the sun shines vertically overhead (in the zenith) at the Tropic of Cancer and Capricorn. The sun never shines in the zenith to the north or south of these respective latitudes.

Tropical Year The period of rotation of the earth about the sun—365.24220 days.

Wot The provincial chief of tribes in southern California.

Zenith The point vertically above the observer in the sky.

Ziggurat A terraced, mud-brick tower erected by the Sumerian and Babylonian civilizations in the Near East.

Zuni The Pueblo village and tribe situated in western New Mexico.

Bibliography

The following works represent a selection of sources which I have found useful in compiling this book. Those marked with an asterisk (*) are of special importance, or will be particularly useful for the general reader seeking to extend his or her understanding of the subject. Anyone wishing to keep up to date with the progress of research in ancient astronomy should consult the *Archaeoastronomy Bulletin*, currently available through John Carlson, the Center for Archaeoastronomy, Space Sciences Building, University of Maryland, College Park, MD 20742, USA.

In the Bibliography the more common periodicals are abbreviated as follows:

AA   American Anthropologist
Am Ant   American Antiquity
Am Sci   American Scientist
Ann NY Acad Sci   Annals of the New York Academy of Sciences
Ann Rev Anthro   Annual Review of Anthropology
Ant   Antiquity
Arch Inst of Am   Archaeological Institute of America
BAE AR   Bureau of American Ethnology Annual Report
BAE Bull   Bureau of American Ethnology Bulletin
BAR   British Archaeological Reports British Series
B Pr Anthro Papers   Ballena Press Anthropological Papers
BSPF   Bulletin de la Societé Préhistorique Française
Bull Soc Polym du Morb   Bulletin de la Societé Polymathique du Morbihan
CA   Current Anthropology
Int Jr Am Ling   International Journal of American Linguistics
Ir Arch Res F   Irish Archaeological Research Forum
J Br Astr Assoc   Journal of the British Astronomical Association
J Calif & GB Anthro   Journal of Californian and Great Basin Anthropology
JHA   Journal for the History of Astronomy
JHA Arch   Journal for the History of Astronomy, Archaeoastronomy Supplement
J Lat Am Fklre   Journal of Latin American Folklore
J Royal Stat Soc   Journal of the Royal Statistical Society
Math Gaz   Mathematical Gazette
Mem Am Anth Assoc   Memoirs of the American Anthropological Association
Mem Am Phil Soc   Memoirs of the American Philosophical Society
Nat Geog Soc Res Rep   National Geographic Society Research Reports

Phil Tr Royal Soc London Philosophical Transactions of the Royal Society of London
Pl Anthro Plains Anthropologist
Proc Int Cong Am Proceedings of the International Congress of Americanists
School Am Res School of American Research
Sci Science
Sci Am Scientific American
Sci & Arch Science and Archaeology
Univ of Cal Pubs in Am Arch & Eth University of California Publications in American Archaeology and Ethnology

GENERAL WORKS ON THE ANCIENT PAST

Clark, J. G. D. 1975. *World Prehistory in New Perspective.* 3rd ed. Cambridge University Press, New York.
Daniel, G. E. 1968. *The First Civilizations.* Crowell, New York, and Thames and Hudson, London.
———. 1981. *A Short History of Archaeology.* Thames and Hudson, London.
Fagan, B. 1977. *People of the Earth.* Little, Brown, Boston.
———. 1978. *In the Beginning.* Little, Brown, Boston. 3rd ed.
———. 1979. *World Prehistory: A Brief Introduction.* Little, Brown, Boston.
Hodges, H. 1970. *Technology in the Ancient World.* Knopf, New York.
Hole, F. and Heizer, R. F., 1977. *Prehistoric Archeology: A Brief Introduction.* Holt, Rinehart and Winston, New York.
Leakey, R. E. F. 1981. *The Making of Mankind.* New York, and London.
———, and Lewin, R., 1977. *Origins.* Macdonald, London, and Dutton, New York.
Pfeiffer, J. E., 1978. *The Emergence of Man.* Harper and Row, New York, 3rd ed.
———. 1977. *The Emergence of Society.* Harper and Row, New York.
———. 1982. *The Creative Explosion.* Harper and Row, New York.
Sherratt, A. 1980. *The Cambridge Encyclopaedia of Archaeology.* Cambridge University Press, New York.

GENERAL WORKS ON ANCIENT ASTRONOMY

Aveni, A. F., ed. 1975. *Archaeoastronomy in Precolumbian America.* University of Texas, Austin.
———. 1977. *Native American Astronomy.* University of Texas.
———. 1981. *Archaeoastronomy,* in *Advances in Archaeological Methods and Theory.* Vol. 4, Academic Press, London and New York.
———. 1982. *Archaeoastronomy in the New World.* Cambridge University Press, New York.
Baity, E. C. 1973. "Archaeoastronomy and Ethnoastronomy So Far." In *CA* Vol. 14, p. 389.
*Brecher, K. and Feirtag, M., eds. *Astronomy of the Ancients.* MIT Press, Cambridge.
Cornell, J. 1981. *The First Stargazers.* Athlone, London.
Culver, R. B., and Ianna, P. A. 1979. *The Gemini Syndrome.* Pachart, Tucson.
Heggie, D. C., ed. 1982. *Archaeoastronomy in the Old World.* Cambridge University Press, New York.

Hodson, F. R., ed. 1974. "The Place of Astronomy in the Ancient World." *Phil Tr Royal Soc London* A 276.

*Krupp, E. C. 1977. *In Search of Ancient Astronomies.* Doubleday, New York.

Lockyer, N. J. 1894. *The Dawn of Astronomy.* London.

———. 1906. *Stonehenge and Other British Stone Monuments Astronomically Considered.* London.

Marshack, A. 1972. *The Roots of Civilization.* McGraw-Hill, New York, and Weidenfeld, London.

McCluskey, S. 1982. "Archaeoastronomy, Ethnoastronomy and the History of Science." In Aveni, A. F., and Urton, G., eds., *Ethnoastronomy and Archaeoastronomy in the American Tropics, Ann NY Acad Sci,* Vol. 385.

Nilsson, M. 1920. *Primitive Time Reckoning.* Lund.

Neugebauer, O. 1952. *The Exact Sciences in Antiquity.* Princeton, Repr. 1969, Dover Books, New York.

Thorpe, I. J. 1981. "Ethnoastronomy: Its Patterns and Archaeological Implications." In Ruggles, C., and Whittle, A., eds., *Astronomy and Society in Britain During the Period 4000–1500 BC,* Oxford, BAR Brit Ser 88.

*Williamson, R., ed. 1981. "Archaeoastronomy in the Americas." *B Pr Anthro Papers* No. 22, Los Altos, California.

ASTRONOMY IN THE OLD WORLD CIVILIZATIONS: BABYLON, EGYPT, GREECE, CHINA

David, A. R. 1980. *Cult of the Sun.* Dent, London.

———. 1982. *The Ancient Egyptians.* Routledge and Kegan Paul, London.

Edwards, I. E. S. 1976. *The Pyramids of Egypt,* revised ed. Penguin, London.

*Hodson, F. R., ed. 1974. "The Place of Astronomy in the Ancient World." *Phil Tr Royal Soc London* A 276. See papers by Aaboe, Needham, Parker, Sachs.

Jacobsen, T. 1976. *The Treasures of Darkness.* Yale University Press, New Haven.

Krupp, E. C. 1980. "Egyptian Astronomy: the Roots of Modern Timekeeping." In *New Scientist,* January 3, p. 24.

Lloyd, G. E. R. 1979. *Magic, Reason and Experience.* Cambridge University Press, New York.

Needham, J. 1959. *Science and Civilization in China,* Vol. 3. Cambridge University Press, New York.

———. 1981. *Science in Traditional China.* Harvard University Press.

*Neugebauer, O. 1952. *The Exact Sciences in Antiquity.* Princeton, Repr. 1969, Dover Books, New York.

Oates, J. 1979. *Babylon.* Thames and Hudson, London.

Pope, M. 1975. *The Story of Archaeological Decipherment.* Thames and Hudson, London.

Roux, G. 1980. *Ancient Iraq,* 2nd ed. Penguin, New York.

Schafer, E., H. 1977. *Pacing the Void.* U. Calif. Press, Berkeley.

Stephenson, F. R. 1982. "Historical Eclipses." In *Sci Am,* October, p. 170.

GENERAL WORKS ON PREHISTORIC EUROPE

Burgess, C. 1980. *The Age of Stonehenge.* Dent, London.

Evans, J. D., Cunliffe, B., and Renfrew, C., eds. 1981. *Antiquity and Man.* Thames and Hudson, London.

Laing, L., and J., 1980. *The Origins of Britain.* Routledge and Kegan Paul, London.

Megaw, J. V. S., and Simpson, D. D. A. 1979. *An Introduction to British Prehistory.* Leicester University Press.

Phillips, P. 1980. *The Prehistory of Europe.* Allen Lane, London.

Piggott, S. 1968. *The Druids.* Thames and Hudson, London.

Powell, T. G. E. 1980. *The Celts*, revised ed. Thames and Hudson, London.

Renfrew, C. 1973. *Before Civilization.* Jonathan Cape, London and Knopf, New York.

Ritchie, G., and Ritchie, A. 1981. *Scotland: Archaeology and Early History.* Thames and Hudson, London.

Twohig, E. S. 1981. *The Megalithic Art of Western Europe.* Oxford University Press, New York.

NEOLITHIC SITES IN THE BRITISH ISLES (CHAPTERS 2–5, PART 5)

*Atkinson, R. J. C. 1960. *Stonehenge.* Repr. 1979, Penguin, New York.

Balfour, M. 1980. *Stonehenge and Its Mysteries.* Scribner, New York.

Burl, A. 1969–70. "The Recumbent Stone Circles of Northeast Scotland." *Proceedings of the Society of Antiquaries of Scotland.* Vol. 102, p. 56.

*———. 1976. *The Stone Circles of the British Isles.* Yale University Press.

———. 1979. *Prehistoric Avebury.* Yale University Press, New Haven.

*———. 1979. *Rings of Stone.* Frances Lincoln, London.

*———. 1981. *Rites of the Gods.* Dent, London.

Daniel, G. E. 1980. "Megalithic Monuments." in *Sci Am*, July, p. 78.

Hadingham, E. 1975. *Circles and Standing Stones.* Heinemann, London, and Walker, New York.

Michell, J. 1982. *Megalithomania.* Thames and Hudson, London, and Cornell University Press, Ithaca, NY.

O'Kelly, M. J. 1982. *Newgrange: Archaeology, Art and Legend.* Thames and Hudson, London.

Ponting, G., and M. 1977. *The Standing Stones of Callanish.* Stornoway.

Royal Commission on Historical Monuments of England. 1979. *Stonehenge and its Environs.* Edinburgh University Press.

Startin, B., and Bradley, R., 1981. "Some Notes on Work Organization and Society in Prehistoric Wessex." In Ruggles, C., and Whittle, A., *Astronomy and Society in Britain During the Period 4000–1500 BC, BAR Br Ser 88.*

Wainwright, G. J. 1975. "Religion and Settlement in Wessex 3000–1700 BC." In Fowler, P. J., ed., *Recent Work in Rural Archaeology.* Rowman and Littlefield, New Jersey.

———. 1979. *Mount Pleasant, Society of Antiquaries Research Committee Report 37.*

——— and Longworth, I. H., 1971. *Durrington Walls, Society of Antiquaries Research Committee Report 29.*

MEGALITHIC ASTRONOMY AND GEOMETRY (CHAPTERS 2–5)

Angell, I. O. 1976. "Stone Circles: Megalithic Mathematics or Neolithic Nonsense?" In *Math Gaz*, Vol 60, p. 189.

———. 1977. "Are Stone Circles Circles?" In *Sci and Arch* no 19, p. 16.

Atkinson, R. J. C. 1966. "Moonshine on Stonehenge." In *Ant* XL, p. 212.

———. 1975. "Megalithic Astronomy: A Prehistorian's Comments." In *JHA* Vol 6, p. 42.

———. 1977. "Interpreting Stonehenge." In *Nature* Vol 265, Jan 6th.

———. 1982. "Aspects of the Archaeoastronomy of Stonehenge." In Heggie, D. C., ed., *Archaeoastronomy in the Old World*, Cambridge University Press.

Brown, P. L. 1976. *Megaliths, Myths and Men.* Blandford, Poole, Dorset, and Taplinger, New York.

Burl, A. 1976. *The Stone Circles of the British Isles.* Yale University Press, New Haven.

———. 1980. "Science or Symbolism: Problems of Archaeo-astronomy." In *Ant* LIV, p. 191.

———. 1982. "Pi in the Sky." In Heggie, D. C., ed., *Archaeoastronomy in the Old World*, Cambridge University Press.

Charrière, G. 1961. "Stonehenge: Rythmes Architecturaux et Orientation." In *BSPF* Vol 58, p. 76.

Colton, R., and Martin, R. L. 1969. "Eclipse Prediction at Stonehenge." In *Nature* Vol 221, March 15, p. 1071.

Ellegård, A. 1981. "Stone Age Science in Britain?" In *CA* Vol. 22 no 2, p. 99.

Fleming, A. 1975. "Megalithic Astronomy: A Prehistorian's View." In *Nature* Vol 255, p. 575.

*Gingerich, O. 1977. "The Basic Astronomy of Stonehenge." In Brecher, K., and Feirtag, M., eds., *Astronomy of the Ancients*, MIT Press, Cambridge.

*Hawkins, G. S. 1963. "Stonehenge Decoded." In *Nature* Vol 200, p. 306.

———. 1964. "Stonehenge: A Neolithic Computer." in *Nature* Vol 202, p. 1258.

———. 1965. *Stonehenge Decoded.* Doubleday, New York.

———. 1968. "Astro-Archaeology." In *Vistas in Astronomy,* ed. A. Beer, Vol. 10, Pergamon Press, Oxford and New York.

———. 1971. "Photogrammetric Survey of Stonehenge and Callanish." *Nat Geogr Soc Res Rep* No 101.

———. 1973. *Beyond Stonehenge.* Harper and Row, New York.

———. forthcoming. *Mindsteps to the Cosmos.*

*Heggie, D. C. 1981. *Megalithic Science.* Thames and Hudson, London.

———. "Megalithic Astronomy: A Review." In *JHA Arch* 3, S 17.

———. 1982. "Megalithic Astronomy: Highlights and Problems." In Heggie, ed., *Archaeoastronomy in the Old World*, Cambridge University Press.

Hogg, A. H. A. 1981. "The Plan of Woodhenge." In *Sci & Arch* no 23, p. 3.

Hoyle, F., 1966. "Speculations on Stonehenge." In *Ant* XL, p. 262.

———. 1977. *On Stonehenge.* W. H. Freeman, San Francisco.

———. 1980. "Stonehenge as an Eclipse Predictor." in *Ant* LIV, p. 44.

MacKie, E. W., 1974. *Archaeological Tests on Supposed Prehistoric Astronomical Sites in Scotland,* in *Phil Tr Royal Soc London* A 276, p. 169.

———. 1977. *Science and Society in Prehistoric Britain.* Elek, London.

———. 1981. "Wise Men in Antiquity?" In Ruggles, C., and Whittle, A., eds., *Astronomy and Society in Britain During the Period 4000 - 1500 BC,* BAR Br Ser 88.

———. 1982. "Implications for Archaeology." In Heggie, D. C., ed., *Archaeoastronomy in the Old World*, Cambridge University Press.

McCreery, T., Hastie, A. J., and Moulds, T. 1982. "Observations at Kintraw." In Heggie, D. C., ed., *Archaeoastronomy in the Old World*, Cambridge University Press.

Newham, C. A. 1972. *The Astronomical Significance of Stonehenge*. Blackburn, Leeds.

Patrick, J. 1974. "Midwinter Sunrise at Newgrange." In *Nature* Vol 249, No 5457, p. 517.

———. 1981. "A Reassessment of the Solar Observatories at Kintraw and Ballochroy." In Ruggles, C., and Whittle, A., eds., *Astronomy and Society in Britain During the Period 4000-1500 BC*, BAR Br Ser 88.

Ritchie, G. 1982. "Archaeology and Astronomy: An Archaeologist's View." In Heggie, D.C., ed., *Archaeoastronomy in the Old World*, Cambridge University Press.

Ruggles, C., 1981. "Prehistoric Astronomy: How Far Did It Go?" In *New Sci* Vol 90, p. 750.

——— and Whittle, A., eds. 1981. *Astronomy and Society in Britain During the Period 4000-1500 BC*, BAR Br Ser 88.

Simpson, D. D. A. 1966-7. "Excavations at Kintraw, Argyll," In *PSAS* Vol 99, p. 54.

Stevenson, R. B. K. 1982. "Kintraw Again." In *Ant* LVI p. 50.

Thom, A. 1954. "The Solar Observatories of Megalithic Man." In *J Br Astr Assoc* Vol 64, No 8, p. 396.

———. 1955. "A Statistical Examination of the Megalithic Sites in Britain." In *J Royal Stat Soc* A 118, p. 275.

———. 1966. "Megalithic Astronomy: Indications in Standing Stones." In Beer, A., ed., *Vistas in Astronomy* Vol 7, p. 1.

———. 1967. *Megalithic Sites in Britain*. Oxford University Press.

———. 1971. *Megalithic Lunar Observatories*. Oxford University Press.

——— and Thom, A.S., 1978. *Megalithic Remains in Britain and Brittany*. Oxford University Press.

———. 1980. "A New Study of All Megalithic Lunar Lines." In *JHA* Vol 2, S 78.

———. 1982. "Statistical and Philosophical Arguments for the Astronomical Significance of Standing Stones." In Heggie, D.C., ed., *Archaeoastronomy in the Old World*, Cambridge University Press.

——— and Thom, A. S. 1974. "Stonehenge" in *JHA* Vol 5, p. 71, and see Vol 6, p. 19.

Wood, J.E. 1978. *Sun, Moon and Standing Stones*. Oxford University Press.

MEGALITHIC SITES IN BRITTANY (CHAPTER 6)

Charrière, G. 1960. "Les Alignements de Menec." In *BSPF* Vol 57, p. 661, and see Vol 59, p. 168 and 313, and Vol 60, p. 562.

Freeman, P. R. 1975. "Carnac Probabilities Corrected." In *JHA* Vol 6, p. 219.

Giot, P. R., and L'Helgouach, J., and Monnier, J. L. 1979. *Préhistoire de la Bretagne*, Ed. Ouest-France, Rennes.

Hadingham, E. 1981. "The Lunar Observatory Hypothesis at Carnac: A Reconsideration." In *Ant* LV, p. 35.

Le Rouzic, Z. 1908. "Tumulus à Dolmen de Er-Grah et le Grand Menhir Brisé." In *Bull Soc Polym du Morb*, Vol 46, p. 57.

Merrit, R. L., and Thom, A. S. 1980. "Le Grand Menhir Brisé." In *Arch J* Vol 137, p. 27.

Patrick, J., and Butler, C. J. 1974. "On the Interpretation of the Carnac Menhirs and Alignments by A. and A. S. Thom." In *Ir Arch Res F* Vol 1 (2), p. 29, and see Vol 3 (1), p. 33.

Roche D. 1969. *Carnac*. Tchou, Paris.

Thom, A., and Thom, A. S. 1971. "The Astronomical Significance of the Large Carnac Menhirs." In *JHA* Vol 2, p. 147; Vol 3, p. 11 and 151; Vol 4, p. 168; Vol 5, p. 30; Vol 7, p. 11.

———. 1974. Reply to Patrick and Butler in *Ir Arch Res F* Vol 1 (2), p. 40.

·———. 1978. *Megalithic Remains in Britain and Brittany.* Oxford University Press.

GENERAL BOOKS ON NATIVE AMERICANS, PAST AND PRESENT

Brotherston, G. 1979. *Image of the New World.* Thames and Hudson, London.

Castile, G. P. 1979. *North American Indians.* McGraw-Hill, New York.

Jennings, J. D. 1974. *Prehistory of North America.* 2nd ed. McGraw-Hill, New York.

———. 1978. *Ancient Native Americans.* W. H. Freeman, San Francisco.

Kehoe, A. B. 1981. *North American Indians.* Prentice-Hall, Englewood Cliffs.

Snow, D. 1976. *The Archaeology of North America.* Thames and Hudson, London.

Spencer, R. F., and Jennings, J. D. 1977. *The Native Americans.* Harper and Row, New York.

Willey, G. R., and Sabloff, J. A. 1974. *A History of American Archaeology.* Thames and Hudson, London, and W. H. Freeman, San Francisco.

TIMEKEEPING AND LUNAR CALENDARS (CHAPTERS 7 AND 8)

·Cope, L. 1919. "Calendars of the Indians North of Mexico." In *Univ of Cal Pubs in Am Arch & Eth* Vol 16 No. 4.

Diószegi, V., and Hoppál, M. 1978. *Shamanism in Siberia.* Akadémiai Kiadó. Budapest.

Eliade, M. 1964. *Shamanism.* Bollingen Series LXXVI, Pantheon, New York.

Hall, E. T. 1959. *The Voices of Time,* in *The Silent Language,* Doubleday, New York.

Hallowell, A. I. 1937. "Temporal Orientation in Western Civilization and in a Preliterate Society." In *AA* Vol 39, p. 647.

Gipper, H., 1977, in Pinxten, R., ed. *Universalism and Relativism in Language and Psychological Thought.* Mouton, The Hague.

Leach, E. R., 1950. "Primitive Calendars." In *Oceania* Vol XX No 4, p. 245.

·Marshack, A. 1972. *The Roots of Civilization.* McGraw, New York, and Weidenfeld, London.

———. 1972. "Cognitive Aspects of Upper Paleolithic Engraving." In *CA* Vol 13 no 3, p. 445, and see Vol 15, p. 328.

Murray, W. B. 1982. "Calendrical Petroglyphs of Northern Mexico." In Aveni, A. F., ed., *Archaeoastronomy in the New World,* Cambridge University Press.

·Nilsson, M. P. 1920. *Primitive Time Reckoning.* Lund.

Stirling, M. W. 1946. "Concepts of the Sun among American Indians." *Smith Inst Ann Rep,* p. 387.

Turton, D., and Ruggles, C. 1978. "Agreeing to Disagree: the Measurement of Duration in a Southwestern Ethiopian Community." In *CA* Vol 19, p. 585.

Whorf, B. L. 1950. "An American Indian Model of the Universe." In *Int Jr Am Ling* Vol 16, p. 67.

———. 1956. *Language, Thought and Reality.* MIT Press, Cambridge, and Chapman and Hall, London.

ASTRONOMY OF THE GREAT PLAINS (CHAPTER 8)

*Chamberlain, V. D. 1982. "When Stars Came Down to Earth." *B Pr Anthro Papers* Los Altos, California.

Eddy, J. 1977. "Medicine Wheels and Plains Indian Astronomy." In Aveni, A. F., ed. *Native American Astronomy.* University of Texas, Austin.

Hyde, G. E. 1974. *The Pawnee Indians.* University of Oklahoma, Norman.

Kehoe, A. B., and Kehoe, T. 1977. "Stones, Solstices and Sun Dance Sites." In *Pl Anthro* Vol 22 No 76 (1).

———. 1979. "Solstice-Aligned Boulder Configurations in Saskatchewan." National Museum of Man, Mercury Series No 48, Ottawa.

Weltfish, G. 1965. *The Lost Universe.* Basic Books, New York.

ASTRONOMY OF NATIVE CALIFORNIANS (CHAPTER 9)

Bean, L., and King, T., eds. 1974. "Social Organization in Native California," in *'Antap, B Pr Anthro Papers* No 2, Socorro, New Mexico.

——— and Blackburn, T. 1976. *Native Californians: A Theoretical Retrospective.* Ballena Press, New Mexico.

*Grant, C. 1965. *The Rock Paintings of the Chumash.* University of California Press, Berkeley.

Hedges, K. 1981. "Winter Solstice Observing Sites. . ." In Williamson, R., ed., *Archaeoastronomy in the Americas, B Pr Anthro Papers* No 22, Los Altos, California.

*Heizer, R. F., ed. 1978. *Handbook of North American Indians, Vol 8, California.* Smithsonian Institution, Washington DC.

Hudson, T., and Blackburn, T. 1977. *The Eye of the Flute.* Santa Barbara Museum of Natural History.

*——— and Underhay, E. 1978. "Crystals in the Sky." *B Pr Anthro Papers* No 10, Socorro, New Mexico.

——— and Lee, G., and Hedges, K. 1979. "Solstice Observers and Observatories in Native California." In *J Calif & Grt Bsn Anthro* Vol 1 (1), p. 39.

——— and Lee, G. 1981. "Function and Purpose of Chumash Rock Art." In *The Masterkey* Volume 55 No 3, p. 92.

NATIVE ASTRONOMERS OF THE NORTH AMERICAN SOUTHWEST (CHAPTER 10)

Cushing, F. H., 1882-3. "My Adventures in Zuni." In *The Century Magazine,* Vol XXV (III), p. 191 and 500, and XXVI (IV), p. 28. Repr. 1941, The Peripatetic Press, Santa Fe.

———. 1920. "Zuni Breadstuff." *Museum of American Indian Notes* No VIII, New York.

———. 1979. ed. J. Green. *Zuni: Selected Writings of F. H. Cushing.* University of Nebraska Press, Lincoln.

Dozier, H. E. P. 1970. *The Pueblo Indians of North America.* Prentice-Hall, Englewood Cliffs.

Ellis, F. H. 1975. "A Thousand Years of the Pueblo Sun-Moon-Star Calendar." In Aveni, A. F., ed., *Archaeoastronomy in Precolumbian America,* University of Texas Press, Austin.

Haile, B. 1947. *Star Lore Among the Navajo.* Museum of Navajo Ceremonial Art, Santa Fe.

James, H. C. 1974. *Pages from Hopi History.* University of Arizona Press, Tucson.

McCluskey, S. 1977. "The Astronomy of the Hopi Indians." In *JHA* Vol 8, p. 174.

———. 1981. "Transformations of the Hopi Calendar." In Williamson, R., ed., *Archaeoastronomy in the Americas,* B Pr Anthro Papers No 22, Los Altos, California.

———. 1982. "Historical Archaeoastronomy: the Hopi Example." In Aveni, A. F., ed., *Archaeoastronomy in the New World,* Cambridge University Press.

*Ortiz, A., ed. 1979. *Handbook of North American Indians, Vol. 9, Southwest,* Smithsonian Institution, Washington DC.

*Page, S., and Page, J. 1982. *Hopi.* Abrams, New York.

Parsons, E. C., ed. 1936. *Hopi Journal of Alexander M. Stephen.* Columbia Univ. Contributions to Anthro. No. 23.

———. 1939. *Pueblo Indian Religion.* University of Chicago Press.

*Stevenson, M. C. 1904. "The Zuni Indians." *BAE AR* 23.

Tedlock, D. 1972. *Finding the Center.* Dial Press, New York.

Titiev, M. 1944. *Old Oraibi.* Papers of Peabody Museum XXII (1).

Tyler, H. A. 1964. *Pueblo Gods and Myths.* University of Oklahoma, Norman.

Zuni, Pueblo of, 1972. *The Zunis: Self-Portrayals,* Albuquerque.

ARCHAEOLOGY AND ANCIENT ASTRONOMY OF THE NORTH AMERICAN SOUTHWEST (CHAPTER 11)

Brandt, J. C., *et al,* 1975. "Possible Rock Art Records of the Crab Nebula Supernova . . ." In Aveni, A. F., ed., *Archaeoastronomy in Precolumbian America,* University of Texas, Austin.

——— and Williamson, R. 1979. "1054 Supernova and Rock Art." In *JHA Arch* 1, S 1.

Hudson, D. T. 1963. "Anasazi Measurement Systems at Chaco Canyon, New Mexico." In *The Kiva* Vol 38 No 72, p. 27.

* Lister, R. H. 1981. *Chaco Canyon: Archaeology and Archaeologists.* University of New Mexico, Albuquerque.

Newman, E. B., Mark, R. K., and Vivian, R. G. 1982. "Anasazi Solar Marker: The Use of a Natural Rockfall." In *Sci* Vol 217, p. 1036.

* Noble, D. G. 1981. *Ancient Ruins of the Southwest.* Northland Press, Flagstaff.

Reyman, J. E. 1976. "Astronomy, Architecture and Adaptation at Pueblo Bonito." In *Sci* Vol 193, p. 957.

Sofaer, A., *et al.* 1979. "A Unique Solar Marking Construct." In *Sci* Vol 206 p. 283.

Vivian, R. G., and Reiter, P. 1960. *The Great Kivas of Chaco Canyon and Their Relationships,* School of American Research Monograph 22, Museum of New Mexico.

Wilcox, D. R., and Shenk, L. O. 1977. *The Architecture of the Casa Grande and its Interpretation,* Arizona State Museum Arch. Series No 115.

Williamson, R., *et al.* 1975. "The Archaeological Record in Chaco Canyon, New Mexico." In Aveni, A.F., ed., *Archaeoastronomy in Precolumbian America.*

*———. 1981. "North America: A Multiplicity of Astronomies." In Williamson, R., ed., *Archaeoastronomy in the Americas.* B Pr Anthro Papers, No. 22, Los Altos, CA.

————. 1982. "Casa Rinconada, a 12th Century Anasazi Kiva." In Aveni, A. F. ed., *Archaeoastronomy in the New World*. Cambridge University Press.

Winter, J. C. 1977. *Hovenweep*. Arch. Rep. No 3, San Jose State University, California.

TROPICAL ASTRONOMY (CHAPTER 12)

* Aveni, A. F., 1981. *"Tropical Archaeoastronomy."* in *Sci* Vol 213, p. 4504.

————. 1981. "Horizon Astronomy in Incaic Cuzco." In Williamson, R., ed., *Archaeoastronomy in the Americas*, B Pr Anthro Papers, No 22, Los Altos, California.

———— and Hartung. 1981. "The Observation of the Sun at the Time of Passage through the Zenith in Mesoamerica." In *JHA Arch* 3, p. S 51.

———— and Urton, G., eds. 1982. *Ethnoastronomy and Archaeoastronomy in the American Tropics*. Ann NY Acad Sci Vol 385.

———— et al, 1982. *"Alta Vista (Chalchihuites) . . ."* In *Am Ant* Vol 47 (2) p. 316.

Flannery, K. V., and Marcus, J. 1976. "Oaxaca and the Zapotec Cosmos." In *Am Sci* Vol 64, p. 374.

Hawkins, G. S., 1974. *Prehistoric Desert Markings in Peru*, Nat Geog Soc Res Rep, 1967 Projects, p. 117.

Isbell, W. H., 1978. "The Prehistoric Ground Drawings of Peru." in *Sci Am* Vol 239 (4), p. 140.

Lévi-Strauss, C. 1964. *Tristes Tropiques*. Criterion Books, New York.

Lewis, D. 1974. "Voyaging Stars." in Hodson, F. R., ed., *The Place of Astronomy in the Ancient World*, Phil Tr Royal Soc London A 276.

————. 1978. *The Voyaging Stars*. Norton, New York.

* Morrison, T. 1978. *Pathways to the Gods*. Harper and Row, New York.

Reiche, M. 1968. *Mystery on the Desert*. Stuttgart.

* Urton, G. 1981. "The Use of Native Cosmologies in Archaeoastronomical Studies." In Williamson, R., ed., *Archaeoastronomy in the Americas*, B Pr Anthro Papers, No 22, Los Altos, California.

————. 1981. *At the Crossroads of the Earth and Sky*. Texas University Press.

Zuidema, R. T. 1964. *The Ceque System of Cuzco*. Brill, Leiden.

————. 1981. "Inca Observations . . ." In Williamson, R., ed., *Archaeoastronomy in the Americas*, B Pr Anthro Papers, No 22, Los Altos, California.

GENERAL BOOKS ON THE ARCHAEOLOGY OF CENTRAL AMERICA

Adams, R. E. W. 1977. *Prehistoric Mesoamerica*. Little Brown, Boston.

Coe, M. D. 1980. *The Maya*, revised ed. Thames and Hudson, London and New York.

Flannery, K. V. 1982. *Maya Subsistence*. Academic Press, New York.

Gyles, A. B., and Sayer, C. 1980. *Of Gods and Men: Mexico and the Mexican Indian*. BBC, London.

Hammond, N. 1982. *Ancient Maya Civilization*. Rutgers University, New Brunswick, NJ.

León-Portilla, M. 1973. *Time and Reality in the Thought of the Maya*. Beacon Press, Boston.

Nicholson, H. B. ed., 1976. "Origins of Religious Art and Iconography." In *Preclassic Mesoamerica*, UCLA, Los Angeles.

Stephens, J. L., 1841. *Incidents of Travel in Central America, Chiapas and Yucatan*. New York, repr. 1969, Dover Books, New York.

————. 1843, *Incidents of Travel in Yucatan.* New York, repr. 1961 Dover Books, New York.

Thompson, J. E. S. 1970. *Maya History and Religion.* University of Oklahoma, Norman.

Weaver, M. P., 1981. *The Aztecs, Maya and Their Predecessors.* 2nd ed., Academic Press, New York and London.

Willey, G. R., 1980. "Towards an Holistic View of Ancient Maya Civilization." in *Man* Vol 15 (2), p. 249.

ANCIENT MAYAN ASTRONOMY (CHAPTERS 13, 15, 16)

Aveni, A. F., ed. 1975. *Archaeoastronomy in Precolumbian America.* University of Texas.

————. 1977. *Native American Astronomy.* University of Texas.

————. 1979. "Venus and the Maya." In *Am Sci* Vol 67, p. 274.

*————. 1980. *Skywatchers of Ancient Mexico.* Texas University.

*————. 1981. "The Maya Region: A Review." In *JHA Arch* No 3, S 1.

————. 1982, ed. *Archaeoastronomy in the New World.* Cambridge University Press.

————. et. al. 1975. "The Caracol Tower at Chichén Itzá: An Ancient Astronomical Observatory?" In *Sci* Vol 188, p. 977.

————. 1978. "The Pecked Cross Symbol in Ancient Mesoamerica." In *Sci* Vol 202, p. 267.

———— and Hartung, H. 1978. "Three Round Towers in the Yucatán Peninsula," in *Interciencia* Vol 3, p. 136.

————. 1982. "Precision In the Layout of Maya Architecture." In Aveni, A. F., and Urton, G., eds., *Ethnoastronomy and Archaeoastronomy in the American Tropics,* Ann NY Acad Sci Vol 385.

Carlson, J. 1976. "A Geomantic Model for the Interpretation of Maya Sites." In *Dumbarton Oaks Conference on Mesoamerican Sites and World Views,* Dumbarton Oaks, Washington DC.

*————. 1981. "Numerology and Astronomy of the Maya." In Williamson, R., ed., *Archaeoastronomy in the Americas,* B Pr Anthro Papers No 22, Los Altos, California.

Hartung, H. 1971. *Die Zeremonialzentren der Maya.* Graz, Austria.

————. 1977. "Ancient Maya Architecture and Planning . . ." In Aveni, A. F. ed., *Native American Astronomy,* University of Texas.

Lamb, W., 1980. "The Sun, Moon and Venus at Uxmal." In *Am Ant* Vol XLV, p. 79.

Ruppert, K., 1935. *The Caracol at Chichén Itzá. Carnegie Inst.* of *Washington,* pub. 454.

ASTRONOMY OF THE MODERN MAYA (CHAPTER 14)

Colby, B. N., and Colby, L. M., 1981. *The Daykeeper.* Harvard University Press.

Gossen, G. H. 1974. *Chamulas In the World of the Sun.* Harvard University Press.

————. 1974. "A Chamula Solar Calendar Board from Chiapas, Mexico." In Hammond, N., ed., *Mesoamerican Archaeology: New Approaches,* University of Texas, and see paper by Marshack, p. 255.

Lincoln, J. S. 1942. *The Maya Calendar of the Ixil of Guatemala, Carnegie Inst. of Washington* Pub 528, vol 7 no 38, p. 97.

Long, R. C. E. 1948. "Observations of the Sun Among the Ixil of Guatemala." In *Carnegie Inst. Notes on Middle America,* Vol 3 (87), p. 214.

Miles, S. W. 1952. "An Analysis of Middle American Calendars." In Tax, S. ed. *Acculturation in the Americas, Proc of XXIX Int. Cong Am*, University of Chicago, p. 273.

Remington, J. A. 1977. "Current Astronomical Practices Among the Maya." in Aveni, A. F., ed., *Native American Astronomy*, University of Texas.

*Tedlock, B. 1982. *Time and the Highland Maya*. University of New Mexico.

Vogt. E. Z. 1969. *Zinacantan: A Maya Community in the Highlands of Chiapas*. Harvard University Press.

RESEARCHES AT PALENQUE (CHAPTER 15)

Rios, E. E., *et. al.* 1980. *Palenque: Esplendor del Arte Maya*. Mexico City.

Robertson, M. G., ed. 1974. *Primera Mesa Redonda de Palenque*, Part 1. Robert Louis Stevenson School, Pebble Beach, California. See papers by Kubler, Matthews and Schele.

———. 1976. *The Art, Iconography and Dynastic History of Palenque*, Part III. Ref. as above, see particularly papers by Carlson, Hartung, Lounsbury and Schele.

———. 1978. *Tercera Mesa Redonda de Palenque*. Ref. as above, see papers by Aveni and Hartung, Schele.

———. 1980. *Third Palenque Round Table, 1978, Vol 5 Part 2*. University of Texas.

Schele, L. 1977. "Palenque: The House of the Dying Sun." in Aveni, A. F., ed., *Native American Astronomy*, University of Texas.

MAYAN INSCRIPTIONS, CODICES AND OTHER WRITTEN SOURCES (CHAPTER 16)

*Brotherston, G. 1979. *Image of the New World*. Thames and Hudson, London.

Carlson, J., forthcoming. "The Grolier Codex . . ." In Brotherston, G. and Aveni, A. F., eds., *Calendars in Mesoamerica and Peru: Native American Computations of Time*, Oxford, BAR.

Coe, M. D. 1973. *The Maya Scribe and His World*. Grolier Club, New York.

*Kelley, D. 1976. *Deciphering the Maya Script*. University of Texas.

Landa, Diego de. 1978. *Yucatan Before and After the Conquest*. tr. by William Gates of *Relación de las Cosas de Yucatán*. Dover Books, New York.

Lounsbury, F. G. 1978, "Maya Numeration, Computation and Calendrical Astronomy." In Gillispie, C. G., ed., *Dictionary of Scientific Biography* Vol 15, supp. 1, p. 759, Scribner, New York.

———. 1982. "Astronimical Knowledge and Its Uses at Bonampak, Mexico." In Aveni, A. F., ed., *Archaeoastronomy in the New World*, Cambridge University Press.

*Marcus, J. 1976. "Origins of Mesoamerican Writing." In *Ann Rev Anthro* 5, p. 35.

Sahagún, Fr. Bernardino de. 1951. *Florentine Codex*, Book 7. ed. A. J. O. Anderson and C. E. Dibble, *School Am Res, Arch Inst of Am Monograph* 14, Santa Fe.

Thompson, J. E. S. 1950. *Maya Hieroglyphic Writing: An Introduction*. Carnegie Inst. of Washington Pub. 589.

———. 1972. "A Commentary on the Dresden Codex." *Mem Am Phil Soc*, Philadelphia.

———. 1972. *Maya Hieroglyphs Without Tears*. British Museum, London.

Sources of Quotations

1. Quoted in Minkowski, H. 1955. "The Tower of Babel: Fact and Fantasy." *Geographical Magazine,* Vol XXVIII, No. 8, December, p. 390.
2. Strabo. *Geography,* 2.26, SVI, I.
3. Lockyer, N. 1906. "Some Questions for Archaeologists." *Nature,* January 18, p. 280.
4. Michell, J. 1977. *Secrets of the Stones.* Penguin, New York, p. 87.
5. Bord, J., and C. 1976. *The Secret Country.* Walker, p. 19.
6. Piggott, S. 1974. "Concluding Remarks." In Hodson, F. R., ed., *The Place of Astronomy in the Ancient World, Phil Tr Royal Soc London* A 276, p. 276.
7. Hawkes, J. 1967. "God In the Machine." in *Ant* XLI, p. 179.
8. Burl, A. 1976. *Stone Circles of the British Isles.* Yale University Press, p. 22.
9. Burl, A. 1979. *Prehistoric Avebury.* Yale University Press, p. 152.
10. Quoted in Hawkins, G. S., and White, J. 1966. *Stonehenge Decoded.* Doubleday, New York, p. 39.
11 and 12. O'Kelly, C. 1971. *Guide to Newgrange.* John English, Wexford, p. 95.
13. Burl, A. 1976. *Stone Circles of the British Isles.* Yale University Press, p. 199.
14. Burl, A. 1979. *Rings of Stone.* Frances Lincoln, London, p. 36.
15. Thom, A., and A. S. 1978. *Megalithic Remains in Britain and Brittany.* Oxford University Press, p. 182.
16. Miln, J. 1877. *Excavations at Carnac.* Edinburgh, p. 100.
17 and 18. Quoted in Brotherston, G. 1979. *Image of the New World.* Thames and Hudson, London, p. 259.
19. Lawrence, D. H. 1931. *Mornings in Mexico.* New York, p. 58.
20. Hall, E. T. 1959. "The Voices of Time." In *The Silent Language,* Doubleday, New York, p. 125.
21. Gipper, H. 1976. In Pinxten, R., ed., *Universalism and Relativism in Language and Thought.* Mouton, The Hague, p. 226.
22. Parsons, E. C. 1917. "Notes on Zuni." *Mem Am Anth Assoc* Vol IV, No 4, p. 297.
23. Murdoch, J. 1890. "Notes on Counting and Measuring Among the Eskimo of Point Barrow." *AA* Vol III, p. 42.
24 and 25. Quoted in Hallowell, A. I. 1937. "Temporal Orientation in Western Civilization and in a Preliterate Society." In *AA* Vol 39, p. 660.
26. Dunbar, J. B. 1882. "The Pawnee Indians." *Magazine of American History,* Vol 8, p. 743.
27 and 28. Fletcher, A. 1902. "Star Cult Among the Pawnee: A Preliminary Report." In *AA* Vol 4, p. 733.
29. Weltfish, G. 1965. *The Lost Universe.* Basic Books, New York, p. 359.
30. See note 26 above.
31. Blackburn, T. 1963. "A Manuscript Account of the Ventureño Chumash." *University of California Arch Survey Ann Rep,* p. 141.
32. Hudson, T., and Underhay, E. 1978. "Crystals in the Sky." *B Pr Anthro Papers* No 10, p. 147.
33. Baxter, S. 1912. *Harper's Magazine,* p. 74.

34. The Pueblo of Zuni, 1973. *The Zunis: Experiences and Descriptions.* p. 16.
35. Cushing, F. H. 1920. "Zuni Breadstuff." *Museum of American Indian Notes and Monographs.* Vol VIII, p. 610.
36. Cushing. F. H. 1883. "My Adventures in Zuni." *Century Magazine,* p. 38.
37. See note 35, p. 174.
38. See note 36.
39. Stephen, A. M., ca. 1883, quoted in McCluskey, S., 1981. "Transformations of the Hopi Calendar." In Williamson, R., ed., *Archaeoastronomy in the Americas,* B Pr Anthro Papers No 22, Los Altos, California.
40. See note 21, p. 225.
41 and 42. Tyler, H. A. 1964. *Pueblo Gods and Myths.* University of Oklahoma, p. 174.
43. Gregory, H. E. 1916. *The Navajo Country, US Geological Water Supply Paper* 380, Washington DC, p. 104.
44. Parsons, E. C. 1925. "Pueblo Indian Journal." *Mem Am Anthrop Assoc* no 32, p. 75.
45. Titiev, M. 1938. "Dates of Planting at Oraibi." *Museum of Arizona Notes* Vol 11 No 5, p. 42.
46. Tedlock. D. 1972. *Finding the Center.* Dial Press, New York, p. 89.
47. See note 36.
48. Krupp, E. C. 1977. *In Search of Ancient Astronomies.* Doubleday, New York, p. 138.
49. Stirling, M. W., 1942, quoted in Williamson, R., ed., 1981. *Archaeoastronomy in the Americas,* B Pr Anthro Papers No 22, p. 72.
50. White, L. A. 1962. "The Pueblo of Sia, New Mexico." *BAE Bull* 184, p. 183.
51. Quoted in Corney, B. G. 1913–19. *The Conquest and Occupation of Tahiti.* Hakluyt Society Vol 11, p. 285.
52. Grimble, A. 1957. *Return to the Islands,* New York, p. 53.
53. Reichel-Dolmatoff, G. 1978. "The Loom of Life: A Kogi Principle of Integration." In *J Lat Am Fklre,* Vol 4 No 1, p. 5.
54. Lévi-Strauss, C. 1977. *Tristes Tropiques.* Pocket Books, New York, p. 271.
55. Quoted in Aveni, A. F. 1981. "Tropical Archaeoastronomy." In *Sci* 10 July, p. 169.
56. Hawkins, G. S. 1974. "Prehistoric Desert Markings in Peru." *Nat Geog Soc Res Reps,* p. 143.
57. López de Cogolludo, D. 1688. *Historia de Yucatán.* Madrid, p. 176.
58. Stephens, J. L. 1841. *Incidents of Travel in Central America, Chiapas and Yucatan,* Vol II, p. 429, Dover reprint 1969.
59. Stephens, J. L. 1843. *Incidents of Travel in Yucatan,* Vol I, p. 221, Dover reprint 1963.
60. See note 57.
61. Gossen, G. 1974. *Chamulas In the World of the Sun.* Harvard University Press, p. 29.
62. Colby, B. N. 1966–67. *Field Notes.* Organized by Human Area Relations Files, p. 58.
63. La Farge, O., and Byers, D. 1931. *The Year Bearer's People.* Tulane University Press, Middle American Research Series No 3, p. 161.
64. Lhuiller, A. R. 1953. "The Mystery of the Temple of the Inscriptions." In *Archaeology,* Spring, p. 6.

65. Von Däniken, E. 1971. *Chariots of the Gods?* Bantam, New York, p. 100.

66. See note 58, p. 354.

67. Lounsbury, F. 1976. "A Rationale for the Initial Date of the Temple of the Cross at Palenque." in Robertson, M. G. R., ed., *The Art, Iconography, and Dynastic History of Palenque* Part III, Robert Louis Stevenson School, Pebble Beach, p. 211.

68. Landa, D. de, 1937. *Relación de las Cosas de Yucatán.* tr. W. Gates, Dover reprint 1978, p. 82.

69. See 68, p. 13.

70. Sahagún, B. 1953. *Florentine Codex.* tr. Anderson, A. J. O., and Dibble, C. E., *Arch Inst of Am Monograph* 14, Part 8, Book 7, University of Utah, p. 12.

71. Aveni, A. F. 1980. *Skywatchers of Ancient Mexico.* University of Texas, p. 181.

72. Wainwright, G. J. 1975. "Religion and Settlement in Wessex 3000–1700 BC." In Fowler, P. J., ed., *Recent Work in Rural Archaeology,* Rowman and Littlefield, New Jersey, p. 68.

73. MacKie, E. W. 1981. "Wise Men in Antiquity?" In Ruggles, C., and Whittle A., *Astronomy and Society in Britain During the Period 4000–1500 BC, BAR Br Ser* 88, p. 137.

74. See note 73, p. 113.

75. MacKie, E. W. 1977. *Science and Society in Prehistoric Britain.* Elek, London, p. 210.

76. See note 73, p. 141.

77. Thompson, J. E. S. 1932. *The Civilization of the Mayas.* Field Museum of Chicago, p. 13.

78. Thompson, J. E. S. 1966. *The Rise and Fall of Maya Civilization.* University of Oklahoma Press, p. 105.

79. Atkinson, R. J. C. 1977. "Interpreting Stonehenge." in *Nature* Vol 265, January 6, p. 11.

80. Aveni, A. F. 1981. "Archaeoastronomy." In *Advances in Archaeological Method and Theory* Vol 4, Academic Press, p. 37.

81. Quoted in Eberhard, W. 1957. "The Political Function of Astronomy and Astronomers in Han China." In Fairbank, J. K., ed., *Chinese Thought and Institutions,* University of Chicago Press, p. 38.

Photo Credits

Aberdeen Archaeological Surveys, Aubrey Burl, 65. American Museum of Natural History, New York, 93, 97, 111, 128, 130, 164, 217, 243. American Philosophical Society, Philadelphia, 27, 222, 223, 224, 225, 232-233. *Radio Times*, British Broadcasting Corporation, London, 34. British Museum, London, 10, 12, 14, 29, 166 (top). Aubrey Burl, 64, 66 (top). Caisse Nationale des Monuments Historiques, SPADEM, Paris, 86. John Carlson, 212 (top), 240. Von Del Chamberlain, 109, 142, 144, 162-163. Michael D. Coe, 220 (bottom). Crown Copyright, Her Majesty's Government, 46, 53. Henry de Lumley, 85. Documentary Media Resources, Cambridge, 16, 22, 24, 35, 41, 43, 69, 70, 75 (bottom), 80, 81, 95 (bottom), 147, 168, 169, 170, 171, 187 (bottom), 199, 219, 220 (top), 235 (top). Mary Ann Durgin, 203. Duncan Earle and Gary Gossen, 197. Fleischer, J. B., 242. Carlo Gay, 176 (margin). Owen Gingerich, 6, 42, 49, 50, 51 (bottom), 54 (top). Gary Gossen, 194, 200, 201. Rosemary Grimble, courtesy of Routledge, Kegan and Paul Ltd, 167. Ian Hampsher-Monk, 66 (center and bottom). Evan Hadingham, 7 (bottom), 32, 33, 36, 61, 74, 75, 77, 78, 119, 138 (top), 143 (bottom left), 145, 146, 148-149, 151, 156, 157, 160, 178 (bottom), 182, 183, 186, 187 (bottom), 188 (bottom), 189, 191, 193, 211 (right), 212 (bottom), 213, 215, 216, 221, 227, 229. Robert Harding Associates, London, 3, 4. Horst Hartung, 176, 179 (top), 184, 214, 240 (bottom). Gerald S. Hawkins, 174, 175. Ken Hedges, 117 (bottom), 118 (top). Heye Foundation, Museum of the American Indian, New York, 100 (right). Chris Jennings, 58. Karl Kernberger, Solstice Project, 153 (bottom), 154. P. Leggate, 235 (bottom). Los Angeles County Museum of Natural History, Western History Collection, 7 (top), 137. Euan MacKie, 55. Merseyside County Museum, Liverpool, 95 (top). Metropolitan Museum of Art, New York, 20. Gordon Moir, 72, 73. Ronald W. B. Morris, 71. William Breen Murray, 91. Musée de l'Homme, Paris, 121. Museum Boymans-van Beuningen, Rotterdam, 8-9. NASA, Washington DC, 68 (margin). National Anthropological Archives, Smithsonian Institution, Washington DC, 94, 106, 107, 114, 122, 124, 126, 127, 129, 134, 135, 138 (bottom), 139. Nebraska State Historical Society, Lincoln, 102 (bottom). Joseph Needham, courtesy of Cambridge University Press, 18, 63, 247. Claire O'Kelly, 52 (margin). Okexnon Films, courtesy of Documentary Educational Resources, Watertown, 196. Mark Oliver and William D. Hyder, 112, 113, 118 (bottom). Oriental Institute, University of Chicago, 15 (bottom). Peabody Museum, Harvard University, 205, 218, 226. Gerald and Margaret Ponting, 25, 30-31. Derek de Solla Price, 19. Robert and Ann Preston, 244. Merle Greene Robertson, 206, 211 (left). Royal Commission on Historical Monuments of England, National Monuments Record, London, 237. Jean-Marie Simon, Visions, New York, 248. D. D. A. Simpson, 39, 54 (bottom). David Hurst Thomas, American Museum of Natural History, New York, 92, 93 (margin). David Turton, 104. Vorderasiatisches Museum, Staatliche Museen Zu Berlin, 15 (top and center). Geoffrey Wainwright, 236. Raymond E. White, 172. Peggy Wier, Solstice Project, 153 (top). Ray Williamson, 47, 82-83, 102 (top), 140, 143 (margin).

Text Figure Credits

23, after A. Fakhry 1961, fig. 68. 40 (left), after A. Thom 1967, 65. 40 (right), after I. Angell 1976, fig. 3. 44-45 (top), based on sketches by Jim Lamb for *Fire of Life: the Smithsonian Book of the Sun*, Smithsonian Books, 1981. 44 (bottom), after excavation plan by courtesy of Michael D. Pitts. 48, after Peter Dunham in Aveni 1981, *Advances in Archaeological Method and Theory*, fig. 1.4. 51 (top), 52 (bottom), after Claire O'Kelly 1967. 56, after Euan MacKie 1974, 179. 59, after A. Thom 1954, fig. 3. 68-69 (upper), after F. R. Stephenson, *Scientific American*, 1982, 172. 68-69 (lower), after J. A. Hynek and N. H. Apfel 1972, 17.15 and 17.19. 76, after Thom and Thom 1978, 9.3. 79, after Z. Le Rouzic 1909. 87, after Breuil and St. Périer 1926. 88, after V. Scerbakiwskyj 1926, fig. VIII and IX. 89, after A. L. Bryan 1978, fig. 1. 90, Brooke, A. de C., *A Winter in Lapland and Sweden*, 1827. 100 (left), T. L. McKenney and J. Hall, *History of the Indian Tribes of North America*, 1837-44, Vol. II, p. 225. 101, F. S. Dellenbaugh, *The North Americans of Yesterday*, 1901, p. 374. 103, after J. Mooney 1898. 108, C. Erskine, *Twenty Years Before the Mast*, 1890. 117 (margin), after T. Hudson, G. Lee and K. Hedges 1979. 120-121 (margins), after T. Hudson and E. Underhay 1978, fig. 8. 125, after E. P. Dozier 1970. 132, after A. Ortiz, ed., Smithsonian *Handbook* 1979. 136, after A. M. Stephen 1936, vol. 2 map 4. 150, after plans by National Park Service. 150 (bottom), after Ray Williamson 1982, fig. 1. 155, after A. Sofaer, R. F. Sinclair and L. Doggett 1982, fig. 4. 159, after National Park Service. 166 (bottom), after Dunham in Aveni 1981, *Science*, fig. 2. 178 (top), after Hartung 1981, *Dumbarton Oaks Conference on Mesoamerican Sites and World-Views*, fig. 15. 179 (bottom), after J. Marcus 1976, fig. 3. 181, after Aveni 1978, *Technology Review*, p. 66. 185, after Aveni and Hartung, *Indiana*, 6, fig. 4. 188 (top), after Aveni, *Technology Review*, p. 63. 192, after Hartung in Aveni and Urton (eds.) 1982, fig. 2d. 195, after R. E. W. Adams, *Antiquity* 54 1980, fig. 1. 202, after Aveni, *Technology Review*, p. 62. 207, after A. Villagra in Ruz Lhuiller 1970, fig. 74. 225 (bottom), after J. A. Hynek and N. H. Apfel 1972, 19.7. 228, after Aveni, Gibbs and Hartung 1975, figs. 5 and 6. 238 (top), after G. J. Wainwright and I. H. Longworth 1971, fig. 33. 238 (center and bottom), after W. Britnell 1982, *Current Archaeology*, p. 201. 241, University Museum, University of Pennsylvania, *Tikal: A Handbook of the Ancient Maya Ruins*, 1967, p. 20.

Index

Italicized page numbers denote references to reproductions and captions.

Aberdeenshire, Scotland, 53, 64-66
Abramson, Philip, 66
Acoma Pueblo, 150, 151
Adveevo (site), 88
Alaska, 101
'Alchuklash, 120, 243
Alexander the Great, 4, 16
Algonquians of eastern North America, 93, *94,* 100
Almagest (Ptolemy of Alexandria), 12, 16
Almanacs, 13, 19, 27, 219, 221-26, 230
Altamira (painted cave), 86, 87
Alta Vista (site), 184-85
American Indians. *See* Native Americans
Ammisaduqa, 12
Anasazi culture, 28, 145, 147, *151,* 152, 153, 155-56, 161
Angell, Ian, *40,* 41
Anglo-Saxon Chronicle, 69
Animal images, 87, 174
Annals (Tacitus), 36
'Antap, 116, 120-23, 243
Antikythera mechanism, 18
Antizenith, 173
Apinayé, of Brazil, 169
Arctic, 63, 101
Argentina, 90
Argyll, Scotland, 53, 58, 77
Arithmetic. *See* Mathematics
Art: cave, 85-87, 111-12, 114; Central American, 175, 180, 183-84; Chumash, 111-12, 118-20; Ice Age, 85-88; Mayan, 187, 198-99, 206-18, 221; rock paintings, 111-12, 114, 118-20, 141, 152, 161; of Southwest, *140*-45, 152, 161. *See also* Carvings
Ashurbanipal, 12
Assyrians, 11-12, *14-15*
Astrology, 5, 15-16, 118
Astronomers, ancient: Babylonians, 4-6, 10-19, 23, 248-49; Chinese, 6, 17-18, 247-48; common characteristics of, 7, 13, 24, 245-49; earliest, 84-88; Egyptian, 17, 19-24, 248-49; Greek, 16-17, 19, 248-49; Mayan, 6-7, 186-93, 201-31, 238-42, 246-49; megalithic, 33-81; Old and New World compared, 26-29, 234-49; tropical, 170-85. *See also* names

of individual cultures; *and* Priests, astronomer
Atkinson, R. J. C., 45-*47*, 246
Aubrey Holes, Stonehenge, *44*
Australia, 90
Aveni, Anthony F., 173, 175, 182, 186-88, 223, 227-28, 230, 246
Aztecs, of Mexico, 176, 181, *221-22*

Babylon (city), 3
Babylonians, 4-6, 10-19, 23, 28, 248-49
Ballochroy (site), 57-60, 72
Barber, John, 63
Barbrook (site), 32
Beliefs. *See* Cosmologies; Religion
Bella Coolla, of Northwest Coast, 102
Beltane, 24
Bering Strait, 88, 90
Black Marsh (site), *40*
Blake, William, *35*
Bonampak (site), 217-18, 247
Books, Mayan, 27-28, 93, 219-26, 230, 231, 234, 239
Bord, Janet and Colin, 35
Bowers Cave (site), *116*
Brahe, Tycho, 72
Britain, prehistoric, 25-26, 32, 34-40, 50-73; circular wooden architecture, 235-38; compared with Maya, 238-39, 242; and the Nootka, 234
British Museum, 10, 11, *14*
Brittany, 38, 67, 73-81
Brotherston, Gordon, 93
Brusch, Heinrich, 10
Building J, Monte Albán, 178-79
Bureau of Ethnology, 125
Burial customs, 50-60, 79-80, 158-59, 208-11
Burl, Aubrey, 36-39, 55, 63-67

Cacique, 131, 147
Caesar, Julius, 21, 43, 71
Cajon (site), 147
Calendars, 14-15, *18*-19, 24-25, 29, 47, 56-57, 59, 85-86, 91, 96, 108, 110, 115, 245; Anasazi, 154-55, 159; California, 118; Central American, 176, 179-81, 198-201; disagreements about, 103-5, 108; horizon, 136, 139-

40, 244, 246; moon, 88, 101-5, 108-9; pre-Christian, 28; Pueblo, 132-34, *136,* 139-40
Calendar sticks, *100,* 108, 118, 131
California, 110-23
California Condor, 115
Callanish (site), 25, *30-1*
Camus-an-Stacca (site), *72*
Capac, Huayna, 170
Caracol, the, Chichén Itzá, 226-30, 239-40
Carlson, John, 215, *220*
Carnac, 76-79
Carnmore (site), 57-59
Carvings, 11-12, 14-15, 27, *51,* 71-72, 78, 86-88, 91-92, 108, 141, 152, 154-56, 179-84, 204-19, 221. *See also* Art
Casa Rinconada, Chaco Canyon, 150, 151-52, 159
Catholicism, 194, 196, 199, 201. *See also* Christianity
Celts, 25-26
Central America, 176-231, 246
Ceques, 171, 173-75
"Ceremonial centers," henges and Mayan sites as, 239-40
Ceremonies, 6, 25-26, 116, 121-22, 126, 128-29, 132, 136-40, 149, 197, 200, 203, 247; Mourning, 122; Snake Dance, *134-35;* White Deerskin Dance, 122
Cerro Colorado (site), 183
Cerro El Chapín (site), *180,* 183-84
Chaco Canyon, *125,* 140-45, 147-52, 156-61, 244-45
Chaldaeans, 16
Chamberlain, Von Del, 106, *142*
Chamula, of Mexico, *194*-97, 200-203, 248
Chan-Bahlum, 213-17
Charlemagne, 165
Chetro Ketl, Chaco Canyon, 148, 156
Cheyenne, of Great Plains, 101
Chiapas, Mexico, *7, 194*-95, 197-98, 203, 217-*18*
Chichén Itzá (site), *187, 221,* 226-30, 239, 240
Chilánes, 221
China, 6, 17-18, *63,* 247-48
Chippewa, of Minnesota, 99
Christianity, 24, 37. *See also* Catholicism

Chuckchi, of Siberia, 93
Chumash, of California, 110-23, 160, 165, 243
Church of Santo Domingo, Cuzco, 170
Circle of the Chiefs, The, 106
Clava cairns, 63-64, 242, 245
Clay tablets, 11-12, 14-15
Cochiti, Pueblo of, 131, 147, 161
Codices, Mayan, 27-28, 93, 219-26, 230, 231, 234, 239; Central Mexican, 29, 95, 188
Coe, Michael D., 220
Comets, 119-20
Comanche, of Great Plains, 100
Computer: decoding Stonehenge, 33, 47, 49, 174; and Nazca, 174-75
Constellations, 106, 108, 109, 115, 144
Copán (site), 198-99, 212
Coricancha (site), 170-71, 173
Cornely, Saint, 78
Cosmologies, 93, 95, 107, 114, 118, 168-70, 196-97, 208-11
Counting devices, 72, 86-8, 91-2, 108-9, 118, 131, 165, 180
Crab Nebula, 142-43
Cross Group, Palenque, 214-15
Cults, 160, 243; ancestor, 214; of California, 120-23; cattle, 78-79; Katchina, 7, 128, 139
Cuneiform script, 11
Cup-and-ring marks, 71-72
Cushing, Frank Hamilton, 124-28, 130-31, 133, 141, 144
Cuzco, Peru, 170-72, 174
Cylinder seals, 14, 15

Datura, 115, 119-20
Daykeepers, 199, 200
Dearborn, David, 173
Death, attitudes toward, 23, 50-60, 208, 245. See also Burial customs; Tombs
Decans, 21
Declination, 26
Democracy, and scientific attitude, 17
Dikov, N. N., 88
Divination, 15, 16, 93, 104, 198-200, 219, 221-26, 247
Dol (site), 75
Dowth (site), 51
Dragons, 207-10
Dresden Codex, 27-28, 219, 222-26, 230, 239
Druids, 35, 43
Duibhré, 63
Dun Arnal (site), 56, 57

Dunbar, J. B.. 105, 108
Durkheim, Emile, 97
Durrington Walls (site), 235-38, 242

Earth lodge, 106-108
Earthmaker, 94
Eclipses, 4-5, 11-13, 15, 33-34, 68-71, 73, 81, 118, 165, 222, 238, 249
Egypt, 10, 16-17, 19-24, 181
England. See Britain, prehistoric
Epping, Joseph, 11
Eskimo, 93, 99, 101
Ethiopia, 104, 245
Eudoxus, 19
Evans-Pritchard, E. E., 97
Evening Star, 12, 94, 107, 115, 122, 133, 188, 224-25, 228
Evolution, human, 84

Fairy Stone (megalith), 74-77, 79, 145, 244
Fajada Butte (site), 153-54, 156, 159, 244-45
Farming, 177, 200-201; and Hopi religion, 136-38, 156; and the Inca, 173; invention of, 84, 88, 92, 110, 245; tropical, 241-42
Fejervary Codex, 95
Fertility, 79-80, 138, 222
Fiestas. See Ceremonies
Fisher, Howard, 151
Fletcher, Alice, 105-7
Förstemann, Ernst, 27, 221
Fortingall (site), 39
Fortune-telling. See Divination
Four Corners region, 82, 125, 145, 147, 161
Four Directions. See Space

Galileo, 4
Gaming boards, 181, 247
Garcilaso de la Vega, 170-71
Gatecliff rock shelter, 92
Gaul, 71
Geometry. See Mathematics
Gibbs, Sharon, 227-28
Gipper, Helmut, 97, 134
Gnomons. See Shadow-casting devices
"Golden Age," prehistoric, 34-35
Golden Eagle, 119
Gontzi (site), 88
Gossen, Gary, 195-97, 201
Grant, Campbell, 111-12
Gray, H. St. George, 237
Great Basin, 91-92
Great Plains, 92, 95, 102, 245
Great Pyramid, 21-23, 181

Greeks, 16-17, 19, 248-49
Grolier Codex, 219-20
Grooved Ware, 238
Ground Hog's Day, 25
Guatemala, 27, 28, 176, 194, 198-201, 240-41, 248

Hall, Edward T., 96-97
Halloween, 25
Hammurabi, 12, 17
Hampsher-Monk, Ian, 66
Harrington, John Peabody, 113-14, 115, 116, 119
Hartung, Horst, 178, 186, 188, 190-91, 227-28, 230
Hawkes, Jacquetta, 36
Hawkins, Gerald S., 6, 33-34, 46-47, 49, 67, 174-75
Hawley, Colonel William, 43
Hebrews, 10
Hedges, Ken, 117, 118
Heel Stone, Stonehenge, 42-45, 47-48
Heggie, Douglas, 46
Henges, 236-39, 243
Hieroglyphs, 7, 10-12, 14, 15, 27, 165, 179, 192, 202-3, 216-19, 231, 239, 242
Hill, Robert D., 75
Hillers, Francis K., 127
Hincks, Edward, 11
Hipparchus, 5, 19
Hogg, A. H. A., 237
Hok, Lord, 216
Holly Canyon, 146, 156
Hopewell culture, 243
Hopi, of Arizona, 6, 29, 96-97, 123, 126, 131-40, 150, 156, 160-61, 243-44, 246; map, 132
House of the Magician, Uxmal, 190
Hovenweep (site), 125, 145-47, 156; Castle, 82, 145-46
Hoyle, Fred, 72
Hsi and Hso brothers, 18
Huacas, 171, 173, 175
Hudson, Travis, 113, 116, 119, 151-52
Hunting-and-gathering, 84, 85, 87-88, 90-92, 110, 118, 120
"Hunting marks," 86
Hupa, of California, 122
Hyder, Bill, 119

Ice age, 84-88, 90-92
Imbolc, 24
Incas, of Peru, 170-76
Infrared photography, 201
Intihuatana Stone (site), 172-73
Iraq, 3, 13, 14
Ireland, 25, 50-53, 63

Ixil Maya, of Guatemala, 200, 201
Izapa (site), 180

Jackson, William H., *107, 135*
Jews, 3
Johanson, Donald C., 84
Jura, Scotland, 55-59

Kabotie, Fred, *128*
Kan Xul, Lord, 215
Katchinas, 7, 128; *Heheya, 139*
Kelley, Charles, 183-84
Kerb-cairns, 53-60
Kerlagad (site), *76*
Kills Two, Sam, *102*
Kintraw (site), 53-57, 59-60, 67, 72, 244
Kiowa, of Great Plains, *102*
Kivas, 96-97, 126, 128, 135, 138, 146, 149-52, 156, 158
Kogi, of Colombia, 168-70
Kohl, J. C., *103*
Kosok, Paul, 174
Krupp, Ed, 144
Kwakiutl, of Northwest Coast, 102

Lady-Beast-with-the-Upturned-Snout, 217, 242
Lamb, Weldon, 191-92
Landa, Friar Diego de, 220-21, 240
La Rumorosa (site), *117*
Lascaux (painted cave), 86, 87
Lawrence, D. H., 96
Leakey, Louis and Mary, 84
Leakey, Richard, 84
Leap year correction, 21, 200
Le Placard (site), 86
Lepsius, Karl Richard, 10
Le Rouzic, Zacharie, 74, 79
Lévi-Strauss, Claude, 97, 169
Lighting effects, *97, 117,* 131, 146, *154-56,* 244. *See also* Shadow-casting devices
Lincoln, J. Steward, 201
Loanhead of Daviot (site), 65
Lockyer, Sir Norman, 25-26
Locmariaquer, 74-75, 79
"Long Count" system, 165
López de Cogolludo, Diego, 186, 190
Lounsbury, Floyd, 211, 216-18
Lovat, Lord, 66
Lowie, R. H., *130*
"Lucy," 84
Lughnasa, 24
Luiseño, of California, 115
Lumley, Henry de, 84-85
Lunar. *See* Moon

Machu Picchu (site), *172-73*
McCluskey, Stephen, 132, *151,* 243
MacKie, Euan, 56, 238-39, 242
Madrid Codex, 29
Maes Howe (site), 53-54
Maler, Teobert, 205
Mané-Lud (site), 78
Maps: Alta Vista, *185;* British prehistoric sites, *37;* California sites, *117;* Carnac region, *76;* Chaco Canyon, *159;* earliest New World sites, *88;* Hopi villages, *132;* important sites, *2;* Mayan sites, *195;* North American tribes, *99;* pueblos, *125;* Tikal, *241;* tropical sites, *177*
Marden (site), 235-36, 238
Marduk, 4
Marques de chasse, 86
Marshack, Alexander, 86-88, 92, 201
Masau, 132, 244
Mathematics, 13-15, 17-19, 28, *34,* 40, 41, 203, 246
Matsakia (site), 130-31, 139, 141, 144, 152
Maya, of Central America, 6, 7, 17-18, 24, 27-29, 92, 163, 165, 185-231, 238-43, 246-48; compared with ancient Britain, 234, 238-39, 242; modern, 6, 29, 194-203, 240, 248
Mayapán (site), 229-30
Megaliths, megalithic astronomy, 33-81; Breton, 74-81; defined, 33; mystical responses to, 34-35; standing stones, 25, *33, 72-73,* 74-81; stone circles, 36-38, 64-67; stone rows, *25,* 57-60; tombs, 38, 50-60, 63-64, 78-79; *see also* Stonehenge
Megalithic Yard and geometry, 33, 40
Mérida, Mexico, 186, 229, 241
Mesa Verde, 140, 161
Mesopotamia, 3, 10, 16, 249
Mexico, 28, 29, 93, 95, 165, 176-85, 194-95, 197-98, 203-19, 241, 248
Michell, John, 34
Midnight sun, 63
Miller, Mary, 217
Miller, William C., 143
Miln, James, 78
Mitla (site), *176*
"Moieties," 152, 169-71; 245
Monte Albán, 177-80
Montgaudier (site), 86, 87
Moon, 14-5, 19, 26, 28, 50, 60, 76-78, 80-81, 87, 100, 103, 105, 115, 179-180, 192, 194, 196; counting of months, 96-109, 132-133, 198; movements, 57, 61-73; phases, 14, 19, 62-63, 86; rites, in Scotland, 63-67
Morning Star, 12, 94, 107, 115, 133, 144, 188-89, 224-25, 228, 230
Morrison, Tony, 175
Mountain Spirits, Apache, 94
Mount Pleasant (site), 235-38, 242
Murdoch, John, 101
Murie, James, 105
Murray, W. Breen, 91
Mursi, of Ethiopia, 104-5, 245
Myths, 93; Babylonian, 4; creation, 92-93; Pueblo, 133, 144; sky, 92, 109, 114, 126; star, 109

Nabonassar, 12
Nabopolasser, 4
Nadir. *See* Antizenith
Nanna-Sin, 4
National Geographic Society, 151
Native Americans, 26-29; earliest, 88-90, 92-94; and time, 95-109, 249. *See also* names of individual tribes
Navajo, of Southwest, 100, 126, 145
Navigation techniques, 166-68
Nazca lines, 173-76
Nebaj, Guatemala, 201-2, 243
Nebuchadnezzar, 4
Newgrange (site), 50-54, 58, 63, 77, 242, 244-45
New Mexico, 124-25, 127
"Nine Roads of the Moon," 63
Nohpat (site), 188-89, 192
Nootka, of Northwest Coast, 234-35
Northwest Coast, 70, 102, 234-35
"Notations," lunar, 86
Numbers, 71, 72, 165; Babylonian system, 10, 13, 19, 224, 249; Mayan system, 192-93, 216-17, 221-26, 246
Nunnery, Uxmal, 190-93, 246

Oaxaca, Mexico, 176-80
Observatories, Mayan, 188-90, 201-2, 214-5, 226-30; megalithic, 42-49, 52, 55-60, 64-67, 73, 75-77; Native North American, 107-8, 116-18, 129-31, 135, 141, 144-47, 152-56; tropical, *168,* 170-79, 183-85
O'Kelly, Claire, 51
O'Kelly, Michael J., 50-51
Old Babylonian Period, 12, *14*

Old Kieg (site), 65-66
Olduvai Gorge, Kenya, 84
Omo River, Ethiopia, 104
Oppert, Jules, 11
Orkney Islands, Scotland, 53
Osiris, 23

Paalmul (site), 230
Pacal, Lord, 211-17, 242
Pachacuti, Inca, 171
Pachamama, 173
Paha, 116, 120
Painted Cave (site), 112
Palace of the Governor, Uxmal, 186-90
Palenque (site), 7, 203-19, 246
Pardon, 78
Parsons, Elsie Clews, 98
Pawnee, of Great Plains, 105-9, 245
Pekwin, 129-31, 135-36, 139, 141, 160, 243
Penasco Blanco (site), 141-42, 144, 159
Perthshire, Scotland, 53
Peru, 170-76
Philip II (King of Spain), 176
Picacho peaks (site), 184-85
Piggott, Stuart, 36
Pima, of Arizona, *100*
Pitts, Michael, *44*, 45
Pleiades, *108*, 183
Poeticon Astronomicon, 16
Point Barrow, Alaska, 101
Pole Star, 22, 99
Polynesia, 166-68
Poma de Ayala, Felipé Huamán, *171*
Pottery, Neolithic, 238
Powamu, 138-39
Prediction. *See* Divination
Presa de la Mula (site), 91
Preston, Ann and Robert, *244*
Price, Derek de Solla, 18
Priests, 118, 120, 122, 188, 198-202, 220-23, 227, 238, 243; astronomer-, 118-19, 123, 160, 193, 215, 217, 237-38; rain, 128; snake, *134*; sun, 7, 123, 125-41, 147, 160
Ptolemy of Alexandria, 5, 12-13, 16-17, 19
Pueblo Bonito (site), 148-52, *156*, 245
Pueblos, 6-7, 28, 96, 98, 125-61; map of sites, *125*
Pyramid of the Sun, Teotihuacán, 181, 183
Pyramids, 21-24, 181, 183, 188-90, 204, 207, 210, 229, *240*
Pythagoras, 13, 41

Qagyuhl, of Northwest Coast, *70*
Quarter Days, 25
Quetzalcoatl, 227-28
Quiché Maya, of Guatemala, 197, 199, 203

Rawlinson, Henry, 11
Reiche, Maria, 174
Reichel-Dolmatoff, Gerardo, 168
Religion, 23, 80-81, 87, 92-97, 106, 110, 231, 249; Chumash, 114-17, 122-23; Mayan, 187, 192-200, 203, 208-11, 216-18, 246; pre-Christian, 25, 28; Pueblo, 126, 134-40, 161. *See also* Cosmologies
Remington, Judith, 197
Reyman, Jonathan, 140
Robertson, Merle Greene, 213
Rodríguez Cabrillo, Juan, 110-11
"Roof of voyaging," 165, 167-68
Romans, 10, 21, 24
Rosebud Sioux reservation, *102*
Rosetta Stone, 10-11
Roughting Linn, *71*
Ruggles, Clive, 40, 104-5
Ruz Lhuiller, Alberto, 204

"Sacred Round," 180, 198, 200, 222-23, 230
Sacrifice, human, 38-39, 204, *221*
Sahagún, Bernardino de, 222
Saint-Malo, France, 75
Salteaux, of Canada, 103
Samhain, 24, 25
Samoans, of Pacific, *108*
Sanctuary (site), 237
San Juan Chamula. *See* Chamula
San Marcos Pass, California, *112*
Santa Ynez Mission, California, 111
Schele, Linda, 211
Science, prehistoric astronomy as, 34-35, 38-40, 59, 60, 65, 68, 246-48
Scientific method, 17, 19, 24, 29, 249
Scotland, 25, 38, 40, 53-60, 63-67, 72-73, 76, 77. *See also* Britain
Scribes, 11-13, 15, 27
Sex, and Breton megaliths, 79-80
Shadow-casting devices, 99, 116, 131, 146, 152-6, *168*, 170, 176-79, *244*
Shadowless day, 171, 177, 183
Shalako, 128-9, 136
Shamanism, 92-95, 112, 115, 117, 119-22, 159, 168-69, *197*, 198, 201, 203, 221
Shamash, *14*-15
Sherente, of Brazil, 170

Shetland Islands, Scotland, 63
Shield-Pacal, Lord. *See* Pacal
"Shining One," 25
Shook, Edwin, 201
Shrines, 64; sun, 116, 130, 139-40, 144-45, 147, 152, 159, 161, 171, *244-45*
Siberia, 88, 91, 93
Simpson, Derek D. A., 54
Sioux, of Great Plains, *102-3*
Skidi Band, Pawnee, of Great Plains, 105-9
Sixtus X, Pope, 74
Sky beings, 14-15, 20-21, 106-7, 114-16, 119-20; 209
Sky Coyote, 115-16, 119-20
Skywatchers. *See* Astronomers, ancient
"Slaughter Stone," Stonehenge, *47*
Smyth, Charles Piazzi, 21-22
Sofaer, Anna, 152-55
Solstices, 50, 56, 73, 78, 105, 108, 115-18, 129-30, 135, 139, 141, 154, 168-69, 183, 185, 243, 245; defined, 48
Sosigenes of Alexandria, 21
"Soul journey," 93-94
South America. 169-76
Southwest, North American, 6, 7, 82, 123-61
Soviet Union, 88, 91, 93
Space, 134-35, 173, 196; four directions of, 29, 92, 94-95, 128, 150, 152, 183, 190, 194; six directions of, 133-34
"Speaking of the blood," 200
Stars, 105-9, 115, 142-44, 165, 167, 196; charts and maps, *109*, *118*, 144
Statistics, and astronomical theories, 46, 61, 152
Stela D, Copán, 198-99
Stephen, Alexander M., 136
Stephens, John Lloyd, 187, 189, 214
Stevenson, Colonel James, 126
Stevenson, Matilda (Tilly) Coxe, 126, *129-30*
Stone circles. *See* Megaliths; Stonehenge
Stonehenge, 6-7, 21, 24, 35-36, 42-49, 62, 67, 72, 174, 236, 242, 244, 249
Strabo, 16
Strassmaier, Johann, 10-11
Street of the Dead, Teotihuacán, 182
Strichen (site), 66
Stukeley, William, 43
Sun, 14-15, *21*, 26, 42, 62, 70, 72,

78, 80, 82, 84, 96, 100, 115-16, 120, 129; and death, 50-60; eclipses, 68, 103; and modern Maya, 194-97, 201-2; and moon's cycle, 103; movements, 48, 57, 73; and the Pueblos, 136-41; in tropics, 165, 183-84

"Sun daggers," 147, 154-56, 161, 244-45

Sundials. See Shadow-casting devices

Sun towers, 131, 145-47, 152, 161

Supernatural: and California astronomy, 114-23; and Kogi, 170; and Pueblos, 132, 134-35, 244

Supernovae, 142-43, 145

Symbols, 51, 65, 71-72, 86-88, 94, 103, 108-9, 118, 141, 144, 173, 183, 188-93, 207-18, 245

Tacitus, 36

Tallies. See Calendars; Counting devices

Taos, Pueblo of, 127

Tedlock, Barbara and Dennis, 199

Teeple, John, 202

Tell Asmar (site), 14

Temple of the Cross, Palenque, 214-19, 225, 247

Temple of the Inscriptions, Palenque, 204-6, 210-11, 213, 216

Temple of the Sun: Coricancha, at Cuzco, 170-71; at Palenque, 214

Temple-towers. See Ziggurats

Teotihuacán (site), 181-85, 241

Terra Amata (site), 84-85

Thom, Alexander, 33-34, 38, 40-41, 55, 57, 59, 61-62, 67-68, 70,

72-73, 75-77, 80, 234-35, 238, 242

Thom, Archie, 61, 68, 75-77, 234-35

Thomas, Trudy, 92

Thompson, J. Eric S., 239-42

Thunder Mountain, Zuni, 127, 129

Tikal (site), 240-41

Time: ancient and modern attitudes, 96-101; measuring, 10, 21, 23, 85, 87-88, 97-105, 131, 134, 136, 197-99, 203, 249

Tirawahat, 106

Titiev, Mischa, 139

Tombs, 50-60, 63-64, 78-79, 204-8, 245

Torreon, Machu Picchu, 172-73

Tower of Babel, 4, 8-9

Trigonometry, 19. See also Mathematics

Tropic of Cancer, 183-84

Tropics, 163-231, 241, 248

Tshizunhaukau, 100

Turton, David, 104-5

Tyler, Hamilton, 136-37

Tzompantli, Chichén Itzá, 221

Ukraine, 88

Underhay, Ernest, 114, 119

United Ancient Order of Druids, 43

Unit House, Hovenweep, 146

Ur (site), 3-4

Ushki, Lake (site), 88

Uxmal (site), 186-93, 226, 240, 246

Varela, Andia y, 166

Venerable Bede, 69

Venus, 12, 14, 15, 28, 94, 115, 118, 143-44, 186, 188-93, 197, 215-

16, 218-20, 222, 224-26, 226-31, 242, 247, 249

Village of the Great Kivas (site), 143

Von Däniken, Erich, 208

Vroman, Adam Clark, 7, 137

Wainwright, Geoffrey, 235-37

Walpi, Pueblo of, 135, 136, 139

Weltfish, Gene, 108

Wessex, England, 234, 237

White, Raymond, 173

Whorf, Benjamin Lee, 134

Wijiji (site), 141, 156

Williamson, Ray, 140-41, 145-47, 151, 156

Winnebago, of Great Plains, 100

"Winter Counts," 102-3

"Wobble," lunar, 73, 77

Woodhenge (site), 237, 245

Writing, 10-12, 26, 165, 180, 242; absence of, 24, 165, 249

Xochicalco (site), 177

Yao, Emperor, 18

Yokuts, of California, 122

Yucatán, Mexico, 186-94, 220-21, 226-30, 240

Zac-Kuk, Lady, 212

Zenith, 135, 167, 170, 173, 176, 179-80, 183-85, 248

"Zenith tube," 177, 179

Zia, Pueblo of, 152

Ziggurats, 3-7; of Ur, 3-4

Zodiac, 10, 230, 248

Zuidema, Tom, 173

Zuni, Pueblo of, 7, 98, 123, 125-33, 135-36, 139-40, 143, 152, 156, 160-61, 165, 243